Clay Microstructure

Clay Microstructure

Richard H. Bennett
Matthew H. Hulbert

D. Reidel Publishing Company
A Member of the Kluwer
Academic Publishers Group
Dordrecht/Boston/Lancaster

International Human Resources Development Corporation
Boston/Houston/London

Library of Congress Cataloging in Publication Data

Bennett, Richard Harold, 1939–
Clay microstructure.

Bibliography: p.
Includes index.
1. Clay—Analysis. 2. Soil mechanics. I. Hulbert, Matthew H. II. Title.
QE471.3.B46 1986 552'.5 85-11804
ISBN 0-88746-065-8
[ISBN 90-277-2263-3 D. Reidel]

Published by D. Reidel Publishing Company
P.O. Box 17, 3300 AA Dordrecht, Holland in copublication with IHRDC

Sold and distributed in North America by IHRDC

In all other countries, sold and distributed by Kluwer Academic Publishers Group, P.O. Box 322, 3300 AH Dordrecht, Holland

Geological Sciences Series

Contents

Contents

Acknowledgments

The assistance and time given by numerous colleagues have helped to make the writing of this text possible. Several professionals read an earlier version of the text and gave useful suggestions and comments, which the writers gratefully appreciate. The reviewers included W. D. Keller, R. Pusch, J. K. Mitchell, G. H. Keller, N. R. O'Brien, R. Fass, W. R. Bryant, B. Hamanaka, A. F. Richards, and T. R. McKee.

We give special thanks and appreciation to F. Nastav for drafting most of the figures, proofreading the manuscript, and helping to illustrate graphically the concepts shown in figure 5.1. Much appreciation is given to L. W. Hulbert for drafting figures 5.2 and 5.3, and to G. Merrill for developing a graphical aid for the construction of potential energy diagrams.

The writers appreciate the encouragement given by various colleagues during early research efforts, with special thanks to J.W. Kofoed. We sincerely thank the Eli Lilly Foundation, which supported the library research during 1979 by M. H. Hulbert. Editorial corrections and refinements suggested by IHRDC editors radically improved the text, and for this the writers are extremely appreciative.

1 Introduction

Plan of Review

This review of clay microstructure is aimed at the diverse group of professionals who share an interest in the properties of fine-grained minerals in sediments. During the last several decades, members of this group have included geologists, soil scientists, soil engineers, engineering geologists, and ceramics scientists. More recently, it has included significant numbers of marine geologists and other engineers. Each of the disciplines has developed special techniques for investigating properties of clay sediments that have proven to be fruitful in answering questions of central interest. Knowledge of clay microstructure—the fabric of a sediment and the physico-chemical interactions between its components—is fundamental to all these disciplines (Mitchell 1956; Lambe 1958a; Foster and De 1971). *Clay fabric* refers to the spatial distribution, orientations, and particle-to-particle relations of the solid particles (generally those less than 3.9 μm in size) of sediment. *Physico-chemical interactions* are expressions of the forces between the particles.

In this review, we trace the historical development of understanding clay microstructure by discussing key scientific papers published before 1986 on physico-chemical interactions in fine-grained sediments and on clay fabric. Since the development follows an intricate path, the current view of clay microstructure is summarized. This summary includes a discussion of the present state of knowledge, the observations made so far, and the facts that are now established. A primary objective of this text is to stimulate interest in and awareness of the developmental history of clay microstructure. We emphasize important research advances and review significant dead ends. We believe this will help readers avoid the pitfalls and assist them in refining the questions that arise continually during practical and theoretical research in clay microstructure. We anticipate this approach will aid readers to gain more appreciation for the intrinsic relations between fundamental properties of fine-grained mineral sediments and the fabric and physico-chemistry.

Understanding clay microstructure is important in a diverse array of scientific and technical problems. Ceramicists and geotechnical engineers face many of the same problems associated with microstructure effects on physical properties; petroleum engineers and disposers of chemical and radioactive wastes are both concerned with changes in permeability in response to chemical, thermal, and mechanical stresses; geophysicists are concerned with effects of microstructure on propagation of various types of energy through sediments; and agronomists need to understand relationships between productivity, cultural practices, and soil microstructure. However, knowledge of clay microstructure developed in a given discipline has often been slow to disseminate to scientists and engineers outside of that discipline. This text will have succeeded to the degree that it improves communication of information on clay microstructure among practitioners of all these disciplines.

Soil, Sediment, and Clay Minerals

In addition to developing different techniques of investigation, each of the disciplines involved in the study of fine-grained mineral sediments has developed its own specialized language. Thus a few words on usage are in order. In a general engineering sense, *soil* is any unconsolidated material composed of discrete solid particles and interstitial gas and/or liquids (Sowers and Sowers 1961). More specifically, soil has been described as a *particulate, multiphase system*; "particulate" refers to a system of soil particles (solids), and "multiphase" pertains to a system consisting of the mineral phase (solids) plus a fluid phase (Lambe and Whitman 1969). The mineral phase is often called the *mineral skeleton,* and the fluid phase is called the *pore fluid.* In some cases, soils contain gases. However, soils below the water table and many deep-sea sediments may be totally saturated (i.e., the interparticle voids are completely filled with water). The pore fluid and its constituents, such as dissolved salts and organic compounds, affect the nature of clay mineral surfaces and, therefore, the processes that transmit force at particle contacts.

Sediment, to a geologist, is a deposit formed by the agents of water, wind, or ice. It is produced by chemical, biological, and physical weathering of solid material on the earth's surface. To the geologist, the term *soil* generally denotes the residual materials that accumulate during weathering (Krumbein and Sloss 1953). Both of the synonymous terms *soil* and *sediment* are used in this review, the usage depending upon the engineering or geological context of discussion.

The most active constituents of a soil (or sediment) are the clay minerals; these are basically hydrous aluminosilicates, which have uniquely layered or platelike structures. Technically, clay minerals are classified as phyllosilicates. Clay mineral particles are generally less than 2 μm in size, although some are larger. The most common clay minerals are kaolinite, smectite, illite, and mixed-layer clays. A few less common clays are allophane, halloysite, chlorite, vermiculite and attapulgite. Some clay minerals are tubular or spherical in form as a result of their crystal structure. Further details of clay mineralogy and classification can be found in works by Grim (1940, 1962, 1968) and Grim and Johns (1958). An excellent review of chemical weathering of silicate materials has been written by Keller (1957). Readers are also referred to studies of clay mineral distribution in the ocean basins by Biscaye (1965), Griffin et al. (1968), Rateev et al. (1969), and Lisitzen (1972).

2 *Physico-Chemical Interactions*

Physico-chemical forces are responsible for aggregating clay particles in suspension and for maintaining the masses once they form. In addition to these forces, gravity is important in settling aggregates and consolidating sediment masses. There have been numerous investigations of physico-chemical interactions in clay systems, and interested readers are referred to Mysels (1959), Grimshaw (1971), van Olphen (1977), and Yarin and Cross (1979). In this text the relations between physico-chemistry and clay fabric are treated in considerable detail. It is appropriate to provide first a brief statement and application of the current view of physico-chemical interactions (for an interesting alternative theory, see Larson 1959, 1963). The current view arose from early suggestions that were discarded but rediscovered, false leads that were pursued, and sudden advancements. This complex development of the theory of physico-chemical interactions has been reflected in reports of research on clay-water systems. Although keeping the current theory in mind simplifies review of research on clay fabric, we recognize that the theory surely will be altered as old ideas are refined on the basis of new observations.

Physico-Chemistry

Physico-chemical interactions are essentially electrical in character. They may be enumerated in roughly decreasing order of strength: Born repulsion forces, covalent bonds, electrostatic interactions, hydrogen bonds, and van der Waals forces. (A more quantitative approach is taken in the section "Clay Mineral Structures and Interactions.") Born repulsion forces express the resistance of matter to compression without limit. The repulsion between atoms becomes large as they approach each other at distances approximating atomic dimensions, and the repulsion is nearly independent of the orientations of the approaching atoms. Covalent bonds are attractive, result from the sharing of pairs of electrons by atoms, are effective at small distances between the atomic centers, and are strongly directional. Electrostatic interactions, which include ionic bonds, result from the attraction of unlike charges and the repulsion of like charges; they are effective at even large separations of the charged particles and are omnidirectional. Hydrogen bonds are attractive and link atoms that are relatively rich in electrons to hydrogen atoms. In order to form a hydrogen bond, the hydrogen atom must be bonded covalently to a small, nonmetallic atom (nitrogen, oxygen, or fluorine). Van der Waals forces are attractive, result from electrical asymmetry of atoms or molecules, and are effective over intermediate distances. In the case of molecules with permanent electrical asymmetry (i.e., with permanent electrical dipoles), the van der Waals forces depend upon the relative orientations of the dipoles. In the case of the temporal or dynamic asymmetry which is exhibited by all atoms and molecules, van der Waals forces are omnidirectional on the average over time. They are sometimes called London forces or London–van der Waals forces (see London 1930).

Silicate Minerals

A discussion of how physico-chemical forces can be used to rationalize some of the properties of silicate minerals clarifies the nature of these forces and sets the stage for a review of the physico-chemical determinants of clay microstructure that follows. Bonding between silicon and oxygen in silicates, including the phyllosilicates related to clay minerals, is intermediate in character between covalent bonding and ionic bonding (Grimshaw 1971). The covalent character of the silicon-oxygen bond is manifested by the directionality of the bond: Four oxygen atoms are distributed tetrahedrally about each silicon atom in all stable silicates. The ionic character is indicated by the development, due to electron transfer, of positive charges on the silicon atoms and negative charges on the oxygen atoms. The ionic character is so marked that for many purposes, the silicates may be considered to consist of units of one silicon atom from which four electrons have been removed (i.e., Si^{4+}) and that is surrounded in a regular tetrahedral array by four oxygen atoms that have gained two electrons (O^{2-}).

Although the exact shapes of atoms and ions are difficult to describe, the properties of the Born repulsion force make it possible to approximate the shapes of atoms or ions coming into contact with each other. Owing to the near independence of Born repulsion to the relative orientations of approaching particles, the shape of each seems to be spherical. Since the repulsive force rises extremely rapidly as the distance between the particles decreases, each atom or ion behaves as a hard sphere of fixed radius. Ionic radii have been assigned by analytically determining distances between the centers of ions in various pairs. For all but the smallest ions, the results of various investigators are reasonably consistent (see Desnoyers and Jolicoeur 1964). Taking the ionic radius of the Si^{4+} ion to be 39 pm (0.39 Å) and that of the O^{2-} ion as 140 pm (1.40 Å), it has been shown that the silicon ion is slightly larger than the central cavity formed by four O^{2-} ions placed in contact and with their centers located at the vertices of a regular tetrahedron (Grimshaw 1971). To a first approximation, massive silicates consist of closely packed arrays of spherical oxide ions in which positive ions fill some of the cavities between the oxide ions (fig. 2.1).

Consider a hypothetical experiment in which silicate tetrahedra are brought together to form a massive silicate of sufficient size (e.g., 2 μg) to be weighed with a microbalance. This tiny mass would contain some 10^{16} tetrahedral silicate units, each bearing four negative charges (one Si with four positive charges and four 0, each with two negative charges in each unit). The electrostatic charge of the 2–μg mass would be so great that it would strike a spark to ground through 5 m of air (see Mysels 1959, 296–297). Thus, well before even a small part of this hypothetical silicate could be built, it would dispose of its accumulating negative charge. This approximate calculation illustrates an important principle: Although submicroscopic separation of electrical charges may occur, no significant macroscopic separation of charges exists in stable systems.

Silicates made up of individual tetrahedra packed together conform to this principle by possessing a number of positive ions in the cavities between oxide ions of separate tetrahedra that is sufficient to produce overall electroneutrality. The positive inorganic ions that occur in natural silicate minerals are all small enough to fit within voids between closely packed oxygen ions. These positive ions are held in place by electrostatic interaction, so it is immaterial from what atoms the positive ions are produced. The only requirement is that they be sufficient in number to balance the charge of the tetrahedra. The locations of positive ions in the hexagonally close-packed array of oxide ions depend upon their sizes in comparison with the size of voids in the oxide ion array. For example, positive ions that form olivine minerals are large enough to contact six oxide ions, which surround them in regular octahedral array;

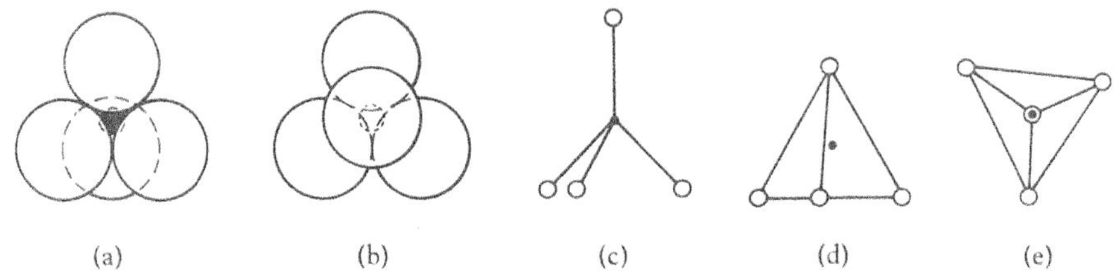

Figure 2.1 *Oxide ions (white) in tetrahedral arrangement about a silicon ion (black): (a) Side view, (b) top views of hard sphere representation; (c,d) side views, and (e) top views of ball-and-stick representations.*

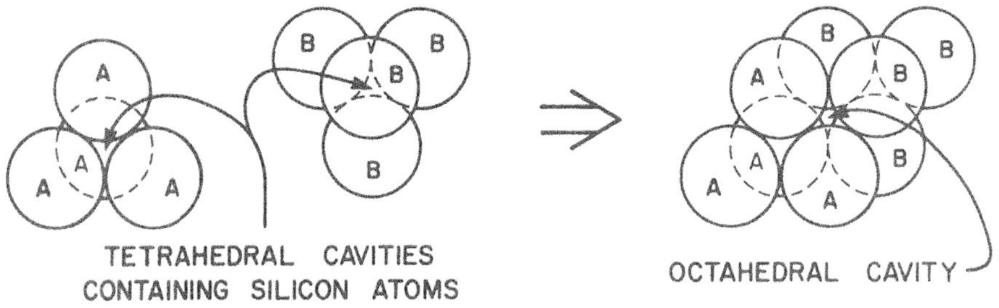

Figure 2.2 *Hard sphere models for olivine group minerals showing two silicate tetrahedra forming a central octahedral cavity, which contains a positive ion. Additional silicate tetrahedra form a crystal of hexagonally close-packed oxide ions. Hidden edges of the projections of spheres are dashed.*

and in the olivine crystal, each one of these positive ions contacts six oxide ions—three from each face of two adjacent tetrahedra (fig. 2.2). The occurrence of natural olivines with all possible ratios of the two charge-neutralizing, positive ions (Mg^{2+} and Fe^{2+}) illustrates the indifference of the crystal to the identity of the cation that stabilizes it electrostatically. This indifference is indicated also by the formation of $CaMgSiO_4$, Mn_2SiO_4, and other silicates with the olivine structure.

The garnet group of minerals also achieves overall electroneutrality by incorporating positive ions into a framework of silicate tetrahedra (Hurlbut 1971). In these complex silicates, positive ions occupy two structurally distinct sites. Rather large, doubly charged (2+) ions occupy dodecahedral sites, whereas smaller, triply charged (3+) ions occupy octahedral sites. As in the case of olivine minerals, a series of minerals is produced as one positive ion is substituted for another similar ion. Thus, natural garnets are found with Fe^{2+} and Mn^{2+} in all possible proportions in the dodecahedral sites, and with Mg^{2+} and Ca^{2+} in these sites in widely varying proportions. Similarly, the octahedral sites in natural garnets may be occupied by Al^{3+}, Fe^{3+}, or Cr^{3+} ions or by some combination of them. Although the identity of ions in either of the types of sites varies, overall electroneutrality is achieved with twelve negative charges from three silicate tetrahedra balanced by six positive charges from three doubly positive ions in dodecahedral sites (x) and six from two triply positive ions in octahedral sites (y). The resulting general formula of the garnet minerals may be expressed as $x_3y_2(SiO_4)_3$.

The excess negative charge of the silicate tetrahedra in a silicate mass may be reduced by each tetrahedron sharing one oxide ion with each of two neighboring tetrahedra, as in the chain silicate structures characteristic of pyroxenes (fig. 2.3). For every silicon ion in the pyroxene structure, only three oxide ions occur and the excess charge is -2 for each tetrahedral unit, rather than -4 as in the olivine structure. Like the olivine crystal, the bulk pyroxene crystal is formed of closely packed oxide ions stabilized electrostatically by the presence of positive ions in the cavities. Again, the identities and even the charges of the extra cations are not critical, as is indicated by the existence of pyroxene minerals with Mg^{2+}, with both Mg^{2+} and Ca^{2+}, and with both Li^+ and Al^{3+} ($MgSiO_3$, $CaMg(SiO_3)_2$, and $LiAl(SiO_3)_2$, respectively). A further reduction

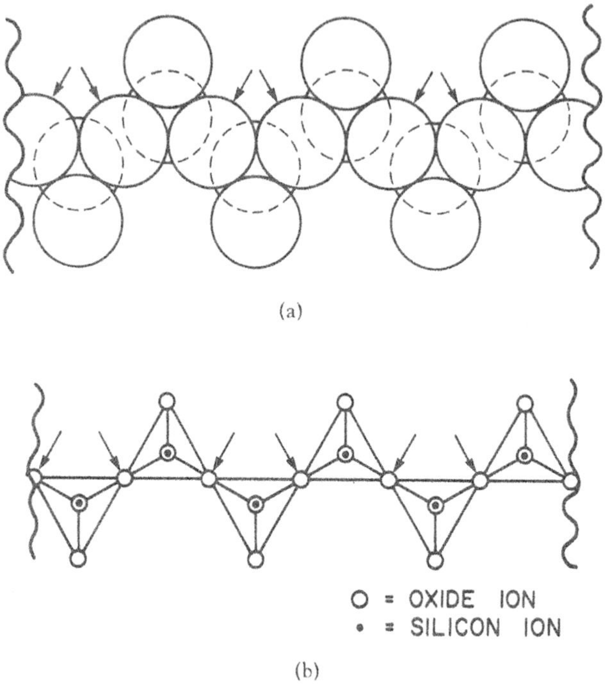

Figure 2.3 (a) *Hard sphere and* (b) *ball-and-stick representations of pyroxene chains. Wavy lines show terminations of drawing of the infinite chains. Arrows point to oxygen atoms that are shared by two tetrahedra.*

in the excess negative charge of each tetrahedral unit exists in the amphibole group of minerals, which are made up of two pyroxene chains linked by the sharing of an oxygen in every alternate tetrahedron (fig. 2.4). If all the oxygens in the plane of the amphibole chain are shared with neighboring chains, the excess negative charge is further reduced, and an infinitely layered structure, characteristic of the micas, is formed. The chains or layers are packed together in the crystals, forming close-packed oxide-ion arrays, and the excess negative charges are balanced by positive ions in the interstices. If every oxide ion is shared by two tetrahedra, a three-dimensional structure is formed, and there no longer occurs excess negative charges associated with the tetrahedral units. Minerals with this structure (e.g., quartz and cristobalite) show no need for compensation of charge by positive ions.

Consideration of the physico-chemical forces can clarify not only the structures of materials (as exemplified by the discussion above), but it also can help to explain the behavior of materials. Consider immersing a piece of quartz in water for a period of years. Little change in the quartz is expected, because the silicon-oxygen bonds are partially covalent and quite strong. Water molecules would not tend to interpose themselves between silicon ions and their neighboring oxygen ions, because much energy is required to increase the distance between these ions, which are held together with bonds of considerable covalent character, and it would not be offset by the energy released by their interaction with the water. The surface of quartz, in contrast to the interior, does provide ions that are available for interactions with water without first requiring energy to break the crystal. Water may be bound to the oxygen ions at the surface of a quartz particle by hydrogen bonds. Since a water molecule possesses a permanent electrical dipole, water molecules very close to the (negative) surface oxygen would spend relatively more time with their positive dipoles oriented toward the quartz as a result of van der Waals interactions (fig. 2.5).

Consider the contrasting behavior of a piece of olivine crystal immersed in water for a period of years. As in quartz, silicon ions in olivine would not be separated from the oxygen ions to which they are bonded because of the large energy required to do so, and the individual silicate tetrahedra would remain intact. At the surface of the mineral, dipolar water molecules would be attracted to the positive ions by van der Waals forces and to the silicate tetrahedra

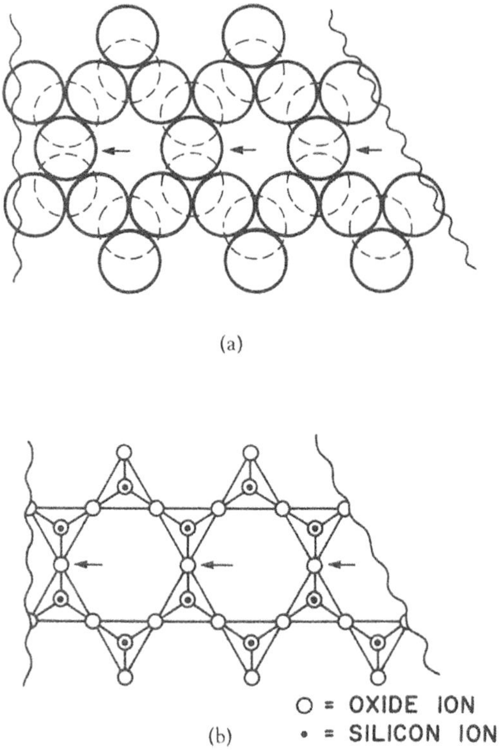

Figure 2.4 (a) *Hard sphere and* (b) *ball-and-stick representations of amphibole double chains. Wavy lines show terminations of drawing of the infinite chains. Arrows point to oxygen atoms that are shared by two pyroxene chains.*

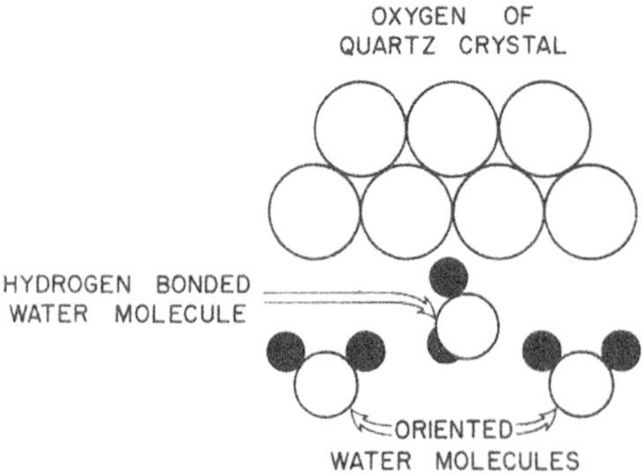

Figure 2.5 *Schematic of quartz particle in water, showing hydrogen-bonded water molecules and water molecules that are preferentially oriented, with the positive end of each dipole oriented toward the quartz surface.*

by both van der Waals forces and hydrogen bonds. The energy released by these interactions at the surface is large enough to overcome electrostatic forces that hold together the positive ions and negative tetrahedra, and as the subunits of the mineral move apart, more surface is exposed for interaction with the water. Thus the olivine crystal breaks apart and dissolves, starting at the surface and proceeding into the interior. Even after dissolution, the positive ions and the negative tetrahedra are separated only by submicroscopic distances, and much of the electrostatic interaction energy remains.

Since the bonding between silicon and oxygen is partially covalent in all silicates, one may conclude that, in general, the silicon-oxygen linkages tend to remain intact in water. Thus pyroxenes and amphiboles should tend to break down to fibers in water, and micas should tend to break down into layers. In these cases, the small positive ions go into solution, and the silicate framework, held together by strong partially covalent bonds, resists solution due to its relatively enormous size. In addition, the ease with which a solid is disrupted by water should decrease with increasing degree of cross-linking of the silicate tetrahedra through oxygen atoms. These conclusions agree entirely with observations of minerals (see the discussion by Goldich [1938] on the responses of minerals to weathering). Although other properties of minerals also may be rationalized in terms of the active physico-chemical forces, the brief treatment above sufficiently sets the stage for discussing the development of hypotheses regarding the physicochemistry of clay microstructure.

Physico-Chemistry of Clays

During the very early days of modern science, long before electrostatic and electrodynamic forces, now termed physico-chemical forces, were understood or even clearly postulated, Reuss (1809) showed that clays suspended in water move toward a positive electrode when a potential is applied (fig. 2.6). That same year, Davy (1809) showed that very finely divided material suspended in water is coagulated by the addition of acid. During 1845, Selmi reported that suspensions of finely divided silver chloride coagulate upon the addition of salts or acids or by stirring (Selmi 1845). He at-

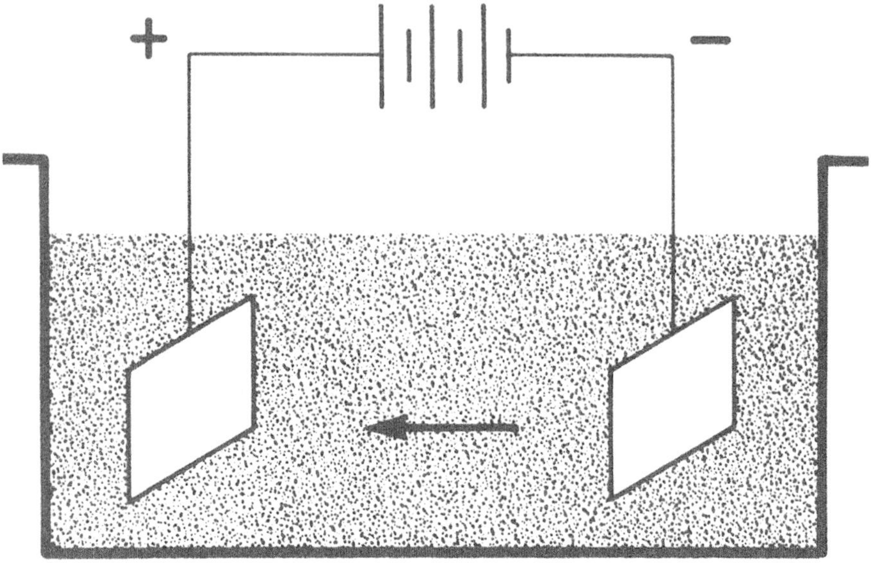

Figure 2.6 *Apparatus for observing effect of an electrical field on the motion of suspended particles in electrolyte solution. Arrow indicates the direction of net transport of suspended clay particles as reported by Reuss (1809).*

tributed the effect of stirring to an increase in the rate of collision between particles, which adhere on contact. Quincke (1861) rediscovered Reuss's work some 50 years after it was first reported; he concluded that particles in contact with a liquid medium accept either a positive or a negative charge, and this charge governs their motion in an electric field. At about the same time, Farady (1857) and Graham (1864) reported their investigations on suspensions of various finely divided materials. These reports laid the groundwork for many of the experimental techniques and much of the nomenclature that are still used in discussing clay-water interaction. Skey (1868) reported that clay in suspension undergoes coagulation and settles out upon the addition of any of a number of neutral salts (e.g., potassium chloride). He proposed that this may be an important phenomenon in nature, causing the oceans to maintain their clarity despite the constant addition of suspended clays by rivers.

The observations by Quincke regarding electrical charges on suspended particles were incorporated into a model that could be treated with mathematical rigor by Helmholtz (1879). Charged particles are assumed to be surrounded by ions of opposite charge at a fixed distance d, and they are held there in solution by electrostatic forces (fig. 2.7). Since d is of the same order of magnitude as molecular size (~ 1 nm), the general principle that no macroscopic separation of charge occurs is not violated. The simple geometry of the particle and its surrounding ions assumed in the Helmholtz model allows straightforward calculation of the electrical potential of the charged particle with respect to its surroundings (see Bockris and Reddy 1970, 826–835, for further discussion). This potential is referred to as the zeta potential ζ and is given by

$$\zeta = 4\pi qd/\epsilon \tag{2.1}$$

for two unit planes, each carrying a charge q and separated by a distance d in a medium of dielectric constant ϵ. The dielectric constant may be obtained experimentally, and the product qd in equation (2.1) also may be calculated, as outlined below, from the results of experimental measurement of the steady motion of particles under the influence of an electric field. Upon the application of a constant electric field, charged particles in suspension reach a limiting velocity when the accelerating force is exactly equal to the

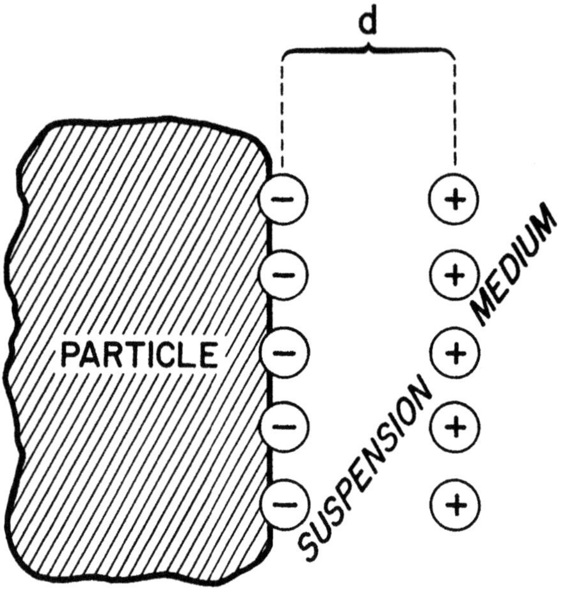

Figure 2.7 *Helmholtz model of a charged particle in suspension, in which oppositely charged layers of ions are separated by distance* d.

retarding viscous drag. The accelerating force is the product of the charge density (q for a particle of unit area) and the electric field strength E. The drag may be expressed in terms of the distance separating the ions from the particle (d) by assuming for simplicity that the relative velocity of the particle with respect to the fluid increases linearly with distance from zero at the particle surface to a constant value at distance d and beyond. In this case, the drag is equal to $\eta v/d$, where η is the viscosity of the fluid, and v is the measured velocity of the particle. Equating the accelerating force and the drag,

$$qE = \eta v/d. \tag{2.2}$$

Solving equation (2.2) for qd and substituting the result into equation (2.1) gives the zeta potential in terms of experimentally accessible quantities:

$$\zeta = 4\pi\eta v/\epsilon E. \tag{2.3}$$

Application of the Helmholtz model to clay particles allows the zeta potential to be used in the quantitative (or more precisely, semiquantitative) description of some properties of clay suspensions and clay sediments saturated with water. Consider clay particles suspended in water. If two like particles are brought close enough together, they will tend to adhere to each other due to the strong, short-range forces that hold the particles together internally. As more and more particles agglomerate, they settle from the suspension and form a sediment in which all the interparticle spaces are filled with water. The tendency of the particles to agglomerate is opposed by the presence of the surrounding layers of ions. For like particles, the sign of the charge of the ion layers is the same (usually positive for clay particles), thus the interaction (electrostatic) between these layers is repulsive. This tends to prevent the close approach of the particles (fig. 2.8). The electrostatic repulsion increases with increasing charge density of the counter-ion layer (q) and decreases with increasing dielectric constant (ϵ) of the medium. Further, the effectiveness of a given degree of repulsion in preventing particle-to-particle contact increases with increasing distance between each particle and its counter-ion layer (increasing d). According to equation (2.1), the zeta potential responds in the

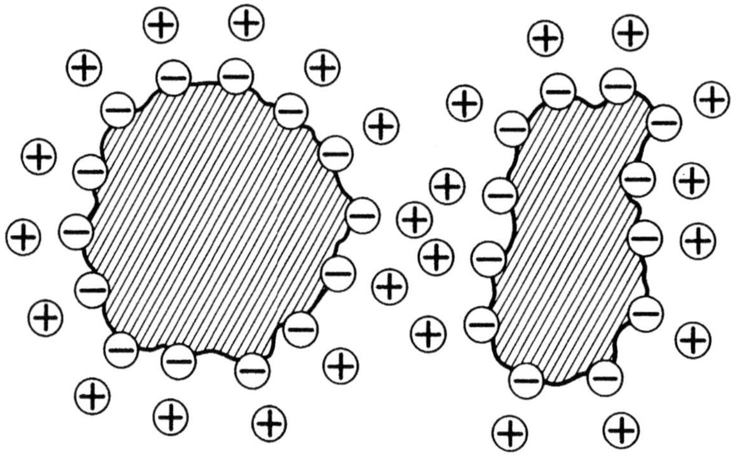

Figure 2.8 *Electrostatic repulsion between the like-charged ion layers tends to prevent coalescence of particles in suspension.*

same way to changes in q, d, and ϵ as does the stability against aggregation. Powis (1916) proposed that a "critical zeta potential" needed to be exceeded in absolute value if a suspension were to be stable. It was suggested that this critical potential is of the order of 20–30 mV.

A large number of investigators have performed experiments in which the stabilities of suspensions and zeta potentials were related to the qualities of electrolytes added to the suspensions. The preeminent investigators were Schulze (1882, 1883), Picton and Linder (1897), and Hardy (1900). Results of all these investigations are summarized below:

1. The ion of primary importance in altering the properties of a suspension is the *counter-ion*, the one with a charge opposite to that of the suspended particle,
2. The effect of this ion in destabilizing suspensions increases rapidly with increasing charge (valence) of the ion,
3. For ions of a given charge, the larger the ionic radius of the bare ion, the greater its destabilizing effect.

The importance of the counter-ion is to be expected if the Helmholtz description is appropriate. Ions of an added electrolyte that have the same charge as the suspended particle would be repelled electrostatically from the particle. Ions of the added electrolyte that have a charge opposite to that of the suspended particle, however, would be present in the counter-ion layer. Because ions move about freely in solution, all the counter-ions (those initially present in solution as well as those of the added electrolyte) would be represented in the counter-ion layer in proportion to their concentration in the solution. The greater effect of more highly charged ions is expected, because less volume would be required to contain enough of these ions to equal the charge of a suspended particle. Accordingly, the thickness, d, of the Helmholtz layer would be reduced, and the repulsive zeta potential would be smaller.

At first glance, the dependence upon size of the counter-ion is the opposite of what might be expected from the Helmholtz model: Smaller ion size should lead to closer approach of the counter-ions to the particle, and consequently, both d and the zeta potential should be smaller. This result was, indeed, puzzling until it was understood that ions in solution do not have the same effective size as bare ions in a vacuum. The effective size of an ion is the volume of the bare ion plus that of any solvent molecules that are more or less permanently attached to it in solution. The smaller the size of an ion, the larger its charge density, and the more strongly it interacts with the solvent. In figure 2.9, the interaction of positive and negative ions with water is shown schematically. Water molecules that are associated with an ion for periods that are long in comparison to the time required for water molecules to become oriented are said to be in the *primary hydration sheath* of the ion (Bockris and Reddy 1970, 45ff). Table 2.1 lists ionic radii (Pauling 1967) and numbers of water molecules in the primary hydration sheaths (Bockris and Reddy 1970) of some ions bearing a single charge. The net result of the interaction between the ions and the solvent is that the sizes of hydrated ions of a given charge are inversely related to the ionic radii of the bare ions. Thus, as is observed, the degree to which the zeta potential and the stability of suspensions should be decreased becomes greater as the radius of the bare ion increases.

These three points that summarize the effects of added electrolytes on the stability of clay suspensions are illustrated by figure 2.10 and 2.11 (see Jenny and Reitemeier 1935). In the system shown in figure 2.11, the critical zeta potential is approximately 42 mV. This figure also shows one important exception to the general rules regarding the effectiveness of added electrolytes: The behavior of the hydrogen ion is anomalous. The effect of monovalent hydrogen ions more nearly approximates the effect of trivalent positive ions. Later investigators (Jackson et al. 1948), in fact, have shown that adding hydrogen ions increases the concentration of the trivalent aluminum ion: Attack on clay minerals by hydrogen ions releases aluminum ions to the solution. Thus the hydrogen ion itself does not determine the effect of the electrolyte; the aluminum ion, which is released from the clay by reaction with a hydrogen ion, does. The Helmholtz model also explains the increase in the tendency of clays to aggregate with increases in the suspension temperature. Small temperature changes alter the Helmholtz layer relatively little, and temperature is not an explicit variable in the function defining the zeta potential (eq. 2.1). However, the increase in thermal agitation of the particles with increased temperature makes it easier for the particles to overcome the repulsion of the counter-ion layers and to approach each other closely.

Table 2.1 *Ionic Radius and Number of Water Molecules in Primary Hydration Sheath for Some Univalent Ions*

Ion	Radius (nm)	Primary Hydration Number
H^+	Vanishingly Small	—
Li^+	0.060	5 ± 1
Na^+	0.095	4 ± 1
K^+	0.133	3 ± 2
F^-	0.136	4 ± 1
Cl^-	0.181	2 ± 1
Br^-	0.195	2 ± 1
I^-	0.216	1 ± 1

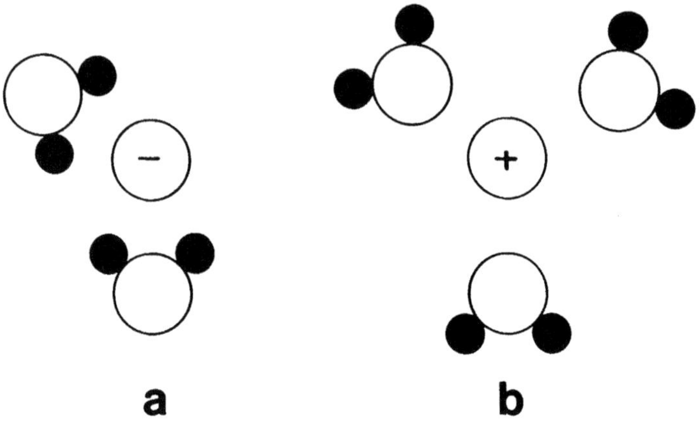

Figure 2.9 *(a) Hydration of negative ion and (b) positive ion in water.*

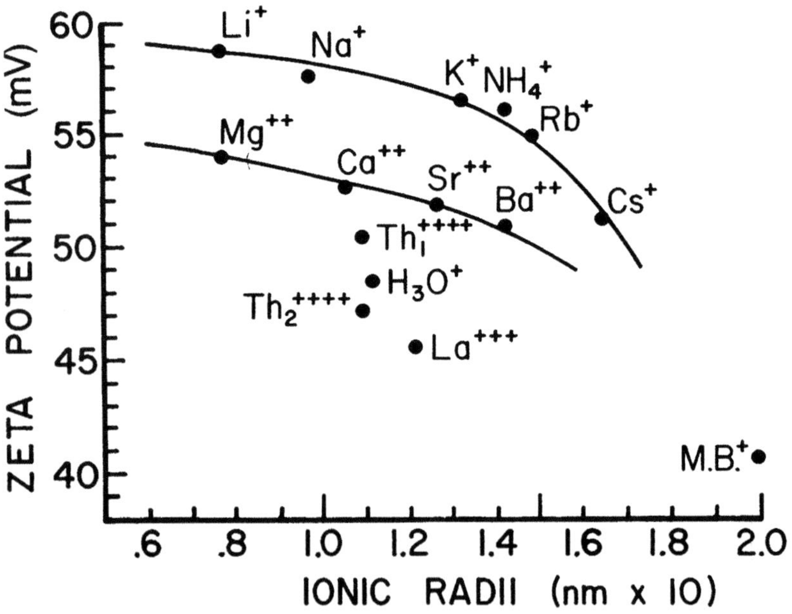

Figure 2.10 *Dependence of the zeta potential of colloidal clay suspensions on charge and ionic radius of counter ions (redrawn from Jenny and Reitemeier 1935). The dot labeled M. B.$^+$ represents cation of methylene blue.*

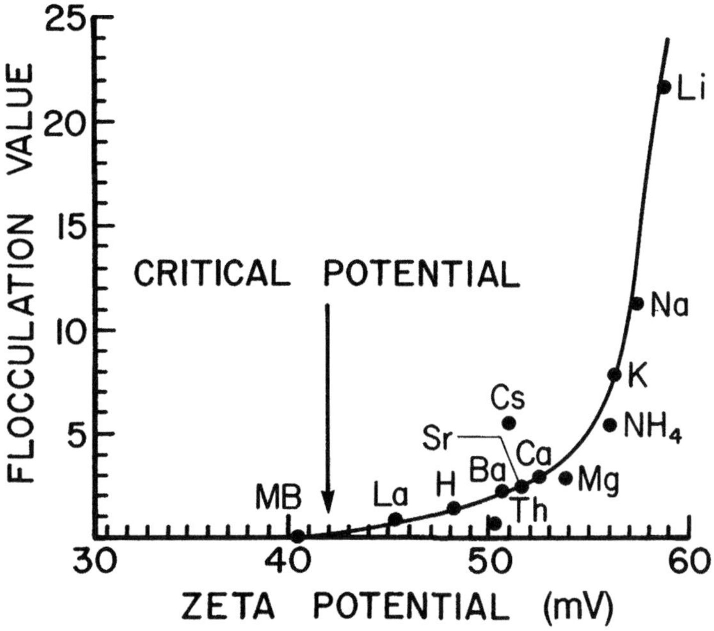

Figure 2.11 *Dependence of the relative amounts of chloride salts required to flocculate clay suspensions (flocculation value) on the zeta potential (redrawn from Jenny and Reitemeier 1935). Identities of counter-ions are given at data points.*

Other Models of Charged Particles

Although the Helmholtz model of charged particles proved to be useful qualitatively, experimental evidence indicates that it is too simple to describe the behavior of suspensions. Gouy (1909) and Chapman (1913) proposed independently that the counter-ions are not held rigidly at a fixed distance from the particle surface, as Helmholtz had suggested, but they diffuse from the surface as a result of collisions with solvent molecules. Stern (1924) synthesized the modern picture of charged particles in contact with a solvent; he combined the Helmholtz and the Gouy-Chapman models. Stern suggested that some counter-ions are held firmly at a fixed distance, close to the surface (the *inner double layer*), and others are held more loosely at a distance from the surface determined by the balance between electrostatic forces and the effects of thermal agitation (the *outer double layer*). Although the results of applying the Stern and Helmholtz models are the same qualitatively, the data required to make rigorous calculations using the Stern model are not obtainable by experiment. The calculations require knowledge of the potential at some known position with respect to the particle. The spatial distribution of potential is determined by the distribution of the electrically charged counter-ions of both the inner and outer double layers. The zeta potential of a particle can be measured, of course, but in the Stern model, the position of the zeta potential with respect to the particle is unknown because the counter-ions of the outer double layer are relatively mobile and diffuse from the surface of the particle. The position of a particle surface is defined satisfactorily, but the magnitude of the potential there is inaccessible experimentally. Therefore, although the Stern model seems to be a faithful representation of charged particles in contact with a solution, its use precludes exact calculation of the electrostatic interactions between particles. Modern models of charged particles in suspension may be regarded as refinements of the Stern model, and possess the same limitations. Some of the refinements may be appreciated by reviewing Outhwaite (1983) and van Oss and Good (1984), and references therein.

Qualitative application of any of the electrostatic models to aggregated clay masses has been somewhat useful for understanding the properties of the sediment. For example, deflocculation of clay masses is made easier by changes that increase the zeta potential of the suspended particles.

Early Ideas of Physico-Chemical Effects on Clay Fabric

Ehrenberg (1918) was among the first to propose a fabric for a clay sediment. At the time of his proposal, the only technique for viewing clay particles involved the use of an ultramicroscope. A beam of light is passed through a suspension approximately perpendicular to the line of sight of the microscope. Light that strikes a particle is scattered, and the part of the light that enters the microscope makes the particle appear as a bright speck against a dark background. Particles that are too small to be seen with a microscope using reflected or transmitted light can be viewed by this technique, but it provides only the most rudimentary indication of their sizes and shapes. Ehrenberg concluded that when electrolyte is added to a clay suspension, a large volume of fluffy sediment forms, because particles adhere on first contact, forming a web or honeycomb structure. He reported that the sediment volume is greater when coagulation is more rapid, because rapid coagulation reduces the opportunity for particles to reorient and thereby reduces the sediment volume. On the basis of similar information, Bradford (1918) came to the opposite conclusion: The sediment does not possess a honeycomb structure, but it is made of stacked spheroidal particles.

Zocher and co-workers (Zocher 1921; Freundlich et al. 1923; Zocher and Jacobson 1928) have investigated the optical properties of concentrated clay suspensions. They concluded that clay particles are shaped like leaves or plates, and are formed into groups of parallel particles. They called a sediment with the particles in parallel array a *Tactoide*. Terzaghi (1925) developed Ehrenberg's proposal and presented a schematic of the fabric of clays deposited in the presence of electrolyte (honeycomb) and in the absence of electrolyte (single-grain structure). Ostwald and Haller (1929) applied Ehrenberg's suggestion by using the sedimentation volume as an indication of the conditions of coagulation (e.g., chemical constitution, temperature, dielectric constant).

Water close to the clay particles was assumed to be important in determining the properties of clay masses (Sorby 1908). As early as 1917, Langmuir had shown that the structures of water and other liquids at surfaces differ from their structures in the bulk liquids (Langmuir 1917). Terzaghi (1920) assumed that water within 50 nm of the surface of clay particles possesses both tensile and shear strength, and clay particles in contact adhere strongly to each other. Goldschmidt (1926) maintained that the rigidity of clay masses results not from the direct contact of clay particles, but from intervening water molecules that adhere strongly to the mineral particles and to each other due to their dipole moments. According to Bancroft (1926), views on the effect of water in clay masses such as those expressed by Terzaghi and Goldschmidt were not accepted generally during the middle 1920s by the community of scholars studying clays. Instead, it was assumed that water surrounds the clay particles in a layer about 50 nm thick and acts as a lubricant between particles. It was also assumed that some material in the water (probably in gelatinous form) bonds the particles together and renders the clay mass plastic. Under these assumptions, the reversibility of moderate drying (i.e., leaving the last few percent of water in the mass) is unexpected, and it was hypothesized that such behavior is governed chemically and caused by the presence of an unknown salt in the aqueous phase.

As pointed out later in the section "Clay Mineral Structures and Interactions," the ideas of a dissolved bonding agent and unknown salt are no longer accepted. However, a special role for and a peculiar character of water very near particles remain fundamental postulates, even though our understanding of them remains quite fragmentary.

The entire field of the colloidal properties of clay systems, until the late 1920s, was characterized by Marshall (1949) as lost in a state of "almost hopeless empiricism." Indeed, in addition to the unsolved problems outlined above, during the late 1920s, the essential nature of the clay particles themselves was in dispute. Bragg et al. (1922) had shown that even submicroscopic particles of kaolinite are crystalline, but investigators disagreed over whether other finely divided clays were crystalline or amorphous. Nevertheless, despite persuasive evidence that Marshall was correct in his pessimistic view of the state of the art, a very important body of experimental evidence and theoretical interpretation existed during the late 1920s. By this time, the electrical nature of the interaction between clay particles in suspension had been recognized, and the essential dependence upon the addition of electrolytes had been reported. A schematic representation of the fabric of clays deposited in the presence and the absence of salts had been presented, and recurring hypotheses of a special function and character for adsorbed water had been advanced.

Clay Mineral Structures and Interactions

The key to renewed progress in understanding the physico-chemical properties of clays was found in the determination of clay crystal structures carried out beginning in 1930 (Pauling 1930a and 1930b). By using X-ray diffraction techniques, the relative positions of atoms in clay particles were determined, and they were shown to be approximately constant in any particular clay mineral. Not only were the clays shown to be crystalline rather than amorphous by this evidence, but also the general shapes and sizes of particles of the various clays were rationalized in terms of physico-chemical forces as was outlined at the beginning of this chapter. It was found that most of the quantitatively important clays consist of assemblages of flat sheets of closely packed oxygen ions, with charge-neutralizing positive ions in the interstices. The charge-neutralizing ions, typically silicon, aluminum, magnesium, potassium or iron ions, balance only approximately the negative oxygen ions. Bonding within sheets is partially covalent and is strong. These sheets differ characteristically according to the identity of the charge-neutralizing ion, and they are stacked commonly into layers in 1:1 or 2:1 combinations. Bonding between layers typically shows little covalent character and is weaker. The layers are held together in manners that are characteristic of the particular clays. In kaolinite, they are held together by hydrogen bonding between hydroxyl groups of one layer and adjacent oxygens of the next layer. In muscovite, the negatively charged layers are held together by electrostatic attraction to potassium ions located between them. In all cases, van der Waals forces are significant in holding the layers together.

The strength of the bonding between layers differs markedly from one clay mineral to another; therefore the responses of various clay minerals to changes in the physico-chemical environment also differ. For example, kaolinite particles remain intact in neutral aqueous medium, because the hydrogen bonds between layers hold them together firmly. In contrast, montmorillonite particles may break down into individual layers when polar water molecules enter between adjacent layers and separate them to such an extent that van der Waals forces are no longer sufficient to keep the particles intact.

When the crystal structures of clay minerals were determined, it became possible to understand the origin of the electrical charges that were known to be critically important in controlling the interactions between clay particles and the interactions of each particle with ions in suspension. Three mechanisms may be seen to be important in establishing a charge on the clay particles: (1) Inexact charge balance within the crystal caused by replacement of ions by others of a different charge and to the finite size of the crystal (i.e., termination of the crystal lattice within unit cells), (2) ionization of surface groups, principally hydroxyls, and (3) adsorption of ions from solution (fig. 2.12). Thus knowledge of the crystal structure aids in the analysis of the interactions of clay particles with each other, solvents, and ions and uncharged substances in solution.

When the structures of clay minerals were being determined, two advances were made in understanding the interactions of finely divided particles; these advances became helpful in studies of clay fabric. Kallmann and Willstaetter (1932) suggested that London forces could be responsible for the attraction between suspended particles, and Hamaker (1936, 1937) developed the graphical method (introduced by Freundlich and Rubin 1935) of drawing curves of potential energy of particle interaction versus distance of separation. In this formulation, the interaction between any pair of clay particles is given by the result of the balance between forces of repulsion (primarily electrostatic, thus related to the zeta potential) and attraction (primarily London–van der Waals). With recognition of the role of attractive forces, the accumulated evidence that the zeta potential alone is insufficient to explain the stability of clay suspensions could be understood.

Figure 2.13 is schematic of potential energy versus interparticle distance for particles of similar charge (for a more extensive discussion of these curves, see Overbeek 1952). Positive potential energies indicate repulsion between the particles; negative potential energies indicate attraction. For particles of like electrical charge, the electrostatic interaction is repulsive and decreases exponentially with increasing distance of separation (curve a, fig. 2.13). The London–van der Waals attraction for planar particles is inversely proportional to the square of the separation distance (curve b, fig. 2.13). As two particles approach each other so closely that their atoms begin to occupy the same space, Born repulsion, which is inversely proportional to approximately the twelfth power of the separation distance, becomes important (curve c, fig. 2.13). The energy of interaction of the particles is the sum of the foregoing potential energies (curve d, fig. 2.13).

For the system shown in figure 2.13, energy would have to be provided in order to move widely separated particles toward each other. If the provided energy is less than the maximum of curve d (at about three units along the distance axis), the particles would separate spontaneously, and the potential energy of the system would decrease in the process. However, once the required energy is available and the particles are brought closer than the distance corresponding to the maximum potential energy, the particles would continue to approach one another with a release of energy until the separation distance is reached corresponding to the minimum of curve d. In order to separate the particles again, energy in an amount equal to the difference between the minimum and the maximum of potential energy shown by curve d would have to be provided. In the system shown in figure 2.13, the amount of energy required to separate the particles is much greater than the energy needed to bring them together. Real clay systems commonly show a similar behavior, with particle separation being more difficult to achieve than particle aggregation. For understanding clay microstructure, this is a major observation.

A slightly more complex case occurs if the potential energy of the system shown in figure 2.13 is recalculated using a somewhat smaller value for the electrostatic repulsion term (fig. 2.14). Here two minima exist, a primary one at about one-half unit of separation and a shallow secondary minimum at about four units of separation. From any separation greater than that of the potential maximum, the particles would move spontaneously to a separation of about four units. If the energy available to the particles from their environment (including thermal energy) is sufficient, the particles may approach more closely than the separation corresponding to the potential maximum, and they will then approach spontaneously to a separation of about one-half unit. Although only a relatively small amount of energy is required to move the particle across the maximum from the secondary minimum at four units, a much greater energy input is required to move the particle pair in the opposite direction from the minimum at one-half unit. A curve for potential energy such as that in figure 2.14 would explain the behavior of systems such as those studied by Zocher and co-workers (see Zocher 1921; Freundlich et al. 1923; Zocher and Jacobson 1928) and briefly discussed above. In these systems, the optical behavior indicates that the clay particles form parallel arrays in suspension, with relatively large separations between par-

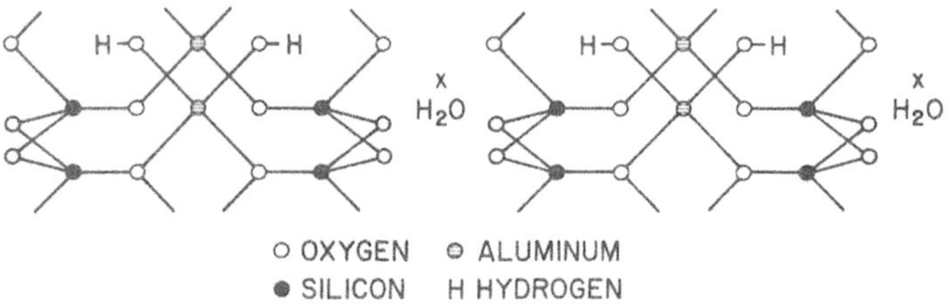

(a)

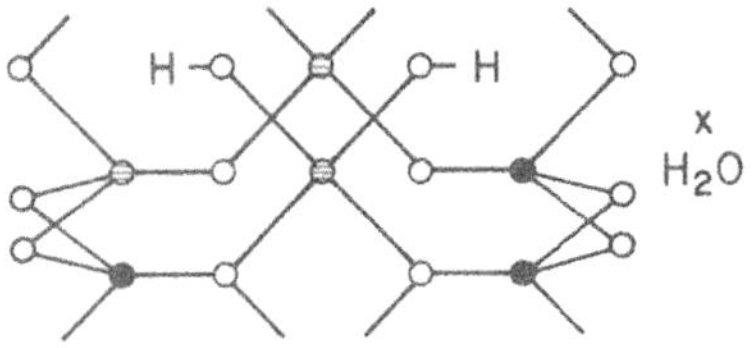

(b)

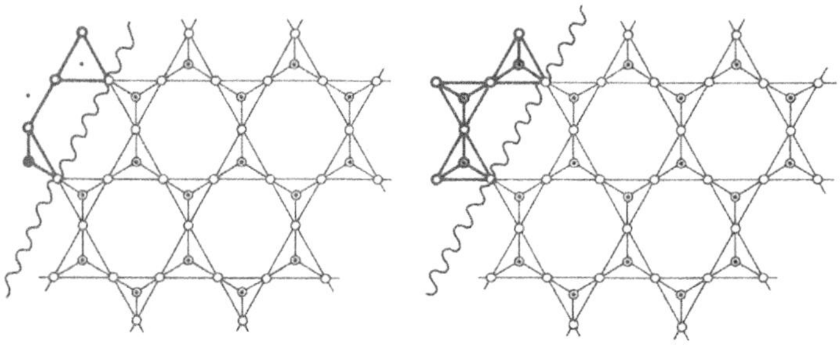

(c)

Figure 2.12 *Development of charge on clay crystals due to replacement of ions of one charge by those of another. (a) A clay crystal consists of unit cells replicated and joined in three dimensions. A hypothetical pair of uncharged unit cells that are structurally similar to those of montmorillonite is shown. (b) Replacement of one Si^{4+} by an Al^{3+} within a unit cell leads to a charge of −1. This charge would be balanced by a positive ion in the interlayer water. (c) Development of charge on clay crystals due to crystal termination. The charge may be positive or negative, and it will depend characteristically on the part of the unit cell that is incomplete. The charges due to the terminating groups of the above infinite sheet silicate are* −6 (left) *and* +2 (right)*; this may be verified by summing the valences of the ions to the right of the wavy termination line. (d) Development of charge by ionization of surface groups. (e) Development of charge on clay crystals due to alteration of ionic groups at the crystal surface. The charge developed due to the loss or grain of ions may be positive or negative.*

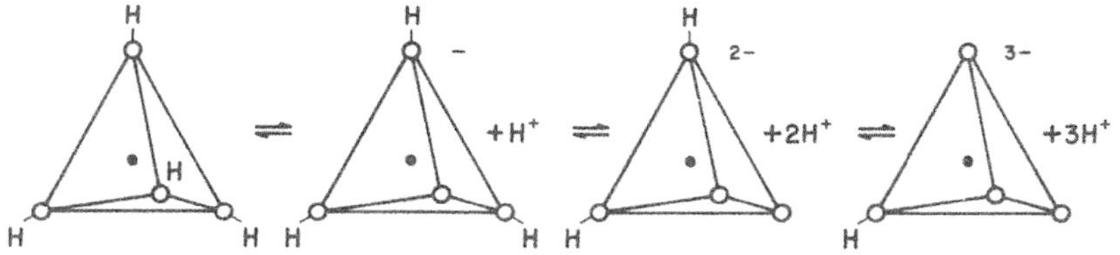

(d)

(e)

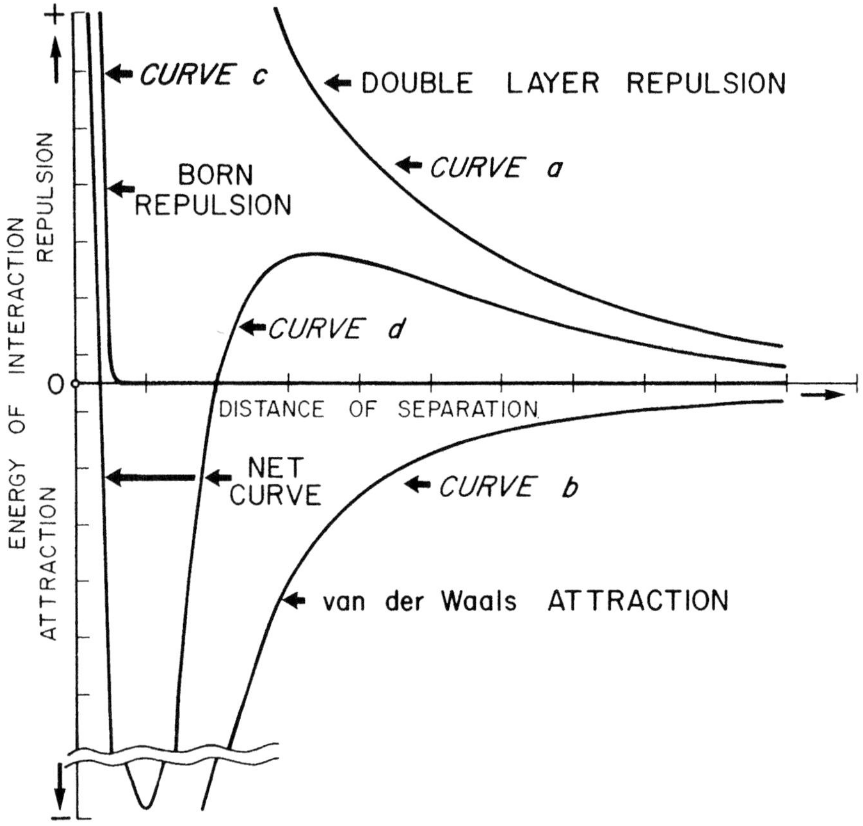

Figure 2.13 *Potential energy versus interparticle distance for two particles in suspension.*

ticles. These arrays can be broken up by gentle mechanical agitation, and they will reform when the suspension is left standing. This behavior might be expected for a system such as that diagrammed in figure 2.14 with separation distances corresponding to about four units along the horizontal axis of this figure. Somewhat more vigorous agitation of the suspensions investigated by Zocher and co-workers led to flocculation that could not be reversed easily; this corresponds to forcing the particles of figure 2.14 past the potential energy maximum to the minimum at one-half unit separation.

Relatively large changes in the potential energies of interaction between particles can be made by small alterations in the composition of the suspension medium. The electrostatic repulsion increases in an exponential fashion with an increase either in the charge density surrounding the particle or in the valence of the ions providing the charge. Repulsion also increases with a decrease in the temperature or the dielectric constant of the medium (Houwink 1938). Such changes produce almost no effect upon the London or Born interaction energies, because these are essentially constant for given particles or minerals. Ionic reactions at the exposed surfaces of particles occur rapidly (Marshall 1964); therefore it might be expected that by altering the electrolyte composition, it would be possible to establish any desired interaction between particles and to alter this interaction at will. Wiegner (1935) observed, however, that the mineral crystal sites to which positive ions bind differ in geometry. The strength of the bonding is a function of the geometry of the site, because covalent bonding possesses a directional character and even purely electrostatic bonding depends upon the closeness of approach. Thus the different sites will exhibit bond strengths of different magnitudes for a given positive ion, and even the relative magnitudes of the bond strengths may not be the same for two different positive ions. Wiegner (1935) concluded that these differences lead to irreversible changes in the system of particles when the electrolyte composition is altered.

Hamaker (1937) pointed out an even more serious problem that arises when the potential energy curves are applied to massive clay sediments. For any specified mass of particles and interstitial fluid, there is a given arrangement of particles that results in the minimum value of potential energy for the system. It is most unlikely, however, that this state of minimum energy would be

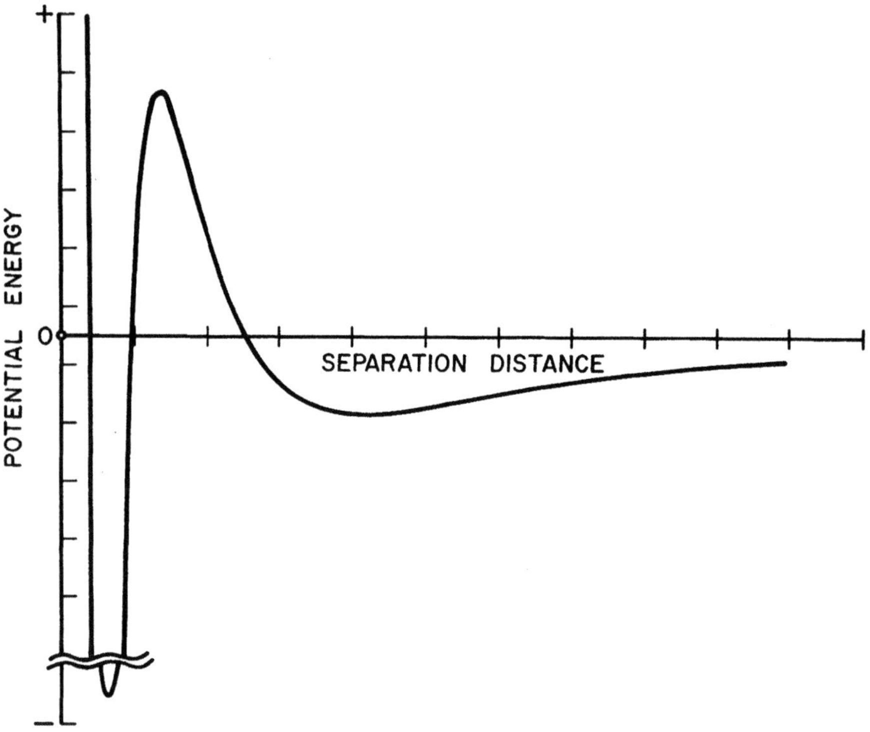

Figure 2.14 *Potential energy versus interparticle distance for two particles in suspension; there are two minima in potential energy.*

reached. The more this state is approached, the more strongly particles will be bound to each other, and the more they will hinder each other in their movements, ultimately making further rearrangement proceed infinitely slowly. Thus, within limits, a set of particle arrangements of nearly minimum energy exists, in which the arrangements are all equally persistent through long periods of time.

Synthesis of Clay Fabric Models

During 1941, Grim summarized the factors that are important in determining the physical properties of clay soil (Grim 1941). He included the kind, amount, and size distribution of the minerals and organic material and the kind and amount of exchangeable cations. He stated that only three groups of clay minerals are generally important—kaolinites, illites, and smectites. The cation exchange capacity (CEC) of these minerals increases in the order given, and Grim suggested that their physical properties may be related to the CEC. Grim (1941) hypothesized that the influence of the various cations is mediated through their effect upon the "hulls" of water surrounding the clay particles. In particular, he observed that water adsorption of the sodium form of montmorillonite was 3 times greater than that of the calcium form. He also reported that a macrofabric of many small fractures developed in montmorillonite upon drying. Hauser and Le Beau (1938) had reported earlier that unoriented gels of bentonite become oriented on drying, forming a film.

During 1946, Rosenqvist (1953) presented striking evidence of sediment chemistry on fabric. He showed that the flocculated structure of certain Scandinavian sediments deposited under marine conditions becomes metastable after much of the salt in the pore water is removed (Rosenqvist 1953). In the absence of mechanical disturbance of the sediment, stability may be recovered through the addition of salt to the pore water. Therefore production of the metasatable materials is reversible in the chemical sense. Application of relatively small mechanical forces to the metastable material leads to structural collapse, often converting the solid system to a freely flowing slurry. Rosenqvist's results are particularly interesting, because the instability of such sediments (often called *quick clays*) has resulted in extensive damage to man-made structures in several localities. Winterkorn (1948) emphasized the importance of the organic component of sediments in controlling fabric; although he specified seven key fractions of the organic matter, he did not define or illustrate the specific effects upon fabric. Skempton and Northey (1952) suggested that leaching of marine clays reduces the thickness of layers of viscous water adsorbed on the particles and thus provides more free water to lubricate the system when particle contacts are broken apart mechanically.

Lambe (1953) summarized the then current concepts of bonding, clay crystallography, and physico-chemical interactions and applied these in order to explain sediment fabric. He stressed use of the surface potential (the potential at the surface referred to the bulk of the material) of the clay particles rather than use of the zeta potential. The potential energy of a pair of particles as a function of the distance between them can be derived from the surface potentials using Gouy-Chapman theory as developed by Derjaguin and Landau (1941) and later by Verwey and Overbeek (1948). The surface potential may be contrasted with the zeta potential: The position of the surface potential with respect to the particle is well defined, but experimental determination of the magnitude of the surface potential presents difficulties. In contrast, the magnitude of the zeta potential is experimentally accessible, but its position with respect to the particle (distance *d* of fig. 2.7) is uncertain (see preceding section). Furthermore, by 1953, a theoretical basis had not been established yet to explain the unexpected but observed dependence of the zeta potential upon the size of the particle. Figure 2.15 (Lambe 1953) shows the sum of the attractive and repulsive potentials for mineral plates as a function of distance between the plates for aqueous solutions of varying concentrations of monovalent salt. Lambe (1953) interpreted these curves to mean that a suspension in the four smallest concentrations of electrolyte will not be flocculated, whereas one at the greatest concentration will flocculate (compare to figs. 2.13 and 2.14). Born repulsion forces, which become significant at very small separations (curve a, fig. 2.13), are neglected in figure 2.15. Thus interparticle distances in the flocculated clays cannot be determined from these curves, and the energy needed to disrupt the flocs cannot be estimated from them.

Rosenqvist (1954) took issue with Lambe's suggestion that the volume of the clay particle controls attraction, and proposed that

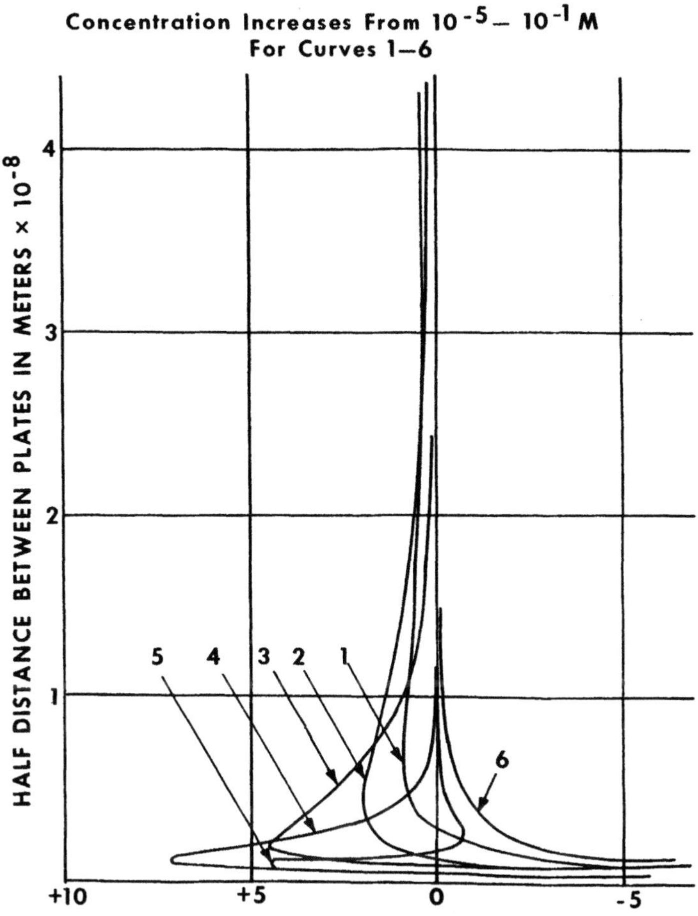

Figure 2.15 *Total potential energy (ergs × 0.1) for two-particle interaction in solutions of various concentrations of electrolyte (redrawn from Lambe 1953). Concentration increases from 0.01 to 100 mM for curves 1 through 6.*

the volume of the adsorbed cation be considered. Rosenqvist suggested also that the dispersant sodium polyphosphate increases particle repulsion by increasing the zeta potential through the adsorption of the polyphosphate anion rather than by increasing the thickness of the adsorbed water layer, as was posited by Lambe. Bjerrum and Rosenqvist (1956) have observed that under saline conditions, the flocculation of silts and clays leads to the deposition of homogeneous sediments of open structure; in fresh water, compact, varved clays of alternating fine and coarse layers are deposited. Mitchell (1956), working in Lambe's laboratory, determined experimentally the orientation of clay particles in thin sections. He concluded that deposition in water of lesser electrolyte concentration leads to local zones of clay particles oriented parallel to each other within a random mass showing abrupt discontinuities in orientation. Sediments deposited under marine conditions or in the presence of significant concentrations of divalent cations were assumed to be initially in a completely random, flocculated state, with positive particle edges bound to negative particle faces. Under mechanical stresses arising from the weight of overlying sediments or from remolding, the flocculated structure breaks down partially to form local areas of parallel particles. Bolt (1956) showed that the particle separation in suspensions of pure homo-ionic clays of grain size less than 0.2 μm could be calculated approximately by taking into account only electrostatic repulsions caused by the various exchangeable cations. This result is not especially surprising, because the attractive van der Waals forces are nearly independent of changes in the solution composition.

Lambe (1958a and 1958b) presented a revision of his discussion of clay fabric and physico-chemical forces. He made particular note of limitations in the assumptions of colloid theory as applied to soil masses: Nonuniformity of particle size and shape, interaction with a number of adjacent particles rather than just one, particle spacing at the lower limit of validity of the theory, and existence of external applied forces, especially those resulting from overlying material. Although Lambe concluded that these limitations prevent general quantitative application of physico-chemical principles to natural sediments, he stated that the principles are, nevertheless, vital to understanding the fundamental behavior. Lambe stated explicitly that water should be considered a "lubricant" for clay particles rather than as a "glue" that holds the particles together. Warentkin and Yong (1960) took exception to the postulate that attraction in clay systems is controlled by van der Waals forces, and they further stated that a force of attraction is not necessarily involved in flocculation. Leonards (1960) suggested that the structure of water in the vicinity of the particles is an important factor in clay structure; this was not accepted by Lambe (1960). Aylmore and Quirk (1959) suggested that polymeric organic molecules may act to bind together domains (regions about 1 μm in size, with clay particles in parallel orientation) into larger stable aggregates.

Rosenqvist (1959) proposed that the interaction between moist clay particles under load differed from that between rock-forming minerals under load, due to the adsorbed water layers about the clay particles. It was proposed that water layers, some 2.5 nm thick, yield plastically as the clay particles are pressed together. In contrast, when rock-forming minerals or dry clays are subjected to loading, elastic stresses are generated at the points of contact. In all materials, short-range van der Waals forces tend to hold the particles together more strongly as the area of contact increases, but if elastic stresses have been produced (such as in rock-forming minerals and dry clays), removal of load leads to release of these stresses and decreases the contact area. Michaels (1961) responded that evidence for solid or very viscous adsorbed water was equivocal, and proposed instead that clay plasticity and cohesion could essentially be explained by using only two assumptions: (1) Spontaneous adhesion between particles occurs if they are brought into sufficiently close proximity, and (2) the geometric arrangement of the particles, which determines the frequency of interparticle contacts, is the primary strength-determining variable. According to Michaels, the initial particle geometry in clay-water systems is determined by the balance of electrostatic repulsion between faces and attraction between edges and faces (Michaels 1961). Electrostatic interactions among the particles are reduced in the dry clay, so application of stress sufficient to break the contacts causes the structure to collapse; the resulting formation of domains (or packets) reduces the contact area. Aylmore and Quirk (1960) concluded that extensive domain formation occurs on drying without remolding, and it persists on rewetting. Dried clays in the calcium form swell on rewetting much less extensively than do clays in the sodium form. Aylmore and Quirk attributed these differences to strong interparticle attraction in the calcium form that prevents reestablishment of the diffuse double layer.

Rosenqvist (1962) modified his proposal regarding adsorbed water at contacts between clay particles and suggested that this water may be in perfect arrangement with no dislocations. By analogy with perfect crystals, such water may yield, but it will not creep. He showed also that replacement of sodium ions by potassium ions in flocculated montmorillonite reduces the thickness of the minerals grains by 18%. He stated that the points of contact between the minerals remain unchanged, but their bonds increase in strength. Martin (1962) and Low (1961) presented reviews of the interaction between clays and water. Both concluded that water in the vicinity of clay surfaces differs in its properties from bulk water. Martin (1962) suggested that the adsorbed water behaves as a two-dimensional fluid, a material that will resist appreciable normal stress but will shear perhaps even more easily than bulk water. Low (1961) concluded that water in the vicinity of the mineral surfaces is structured more rigidly (like a solid). The sharp divergence between the views of these authors is not surprising: Understanding of the structure of pure water and of ionic solutions is still unsatisfactory (see Almagor and Belfort 1978).

In a study of clay fabric using electron microscopy, Smart (1967) concluded that edge contacts are most important for kaolinite, and interparticle bonds derive mainly from broken intracrystalline bonds at the edges of the crystals and from van der Waals forces. Santoro and Stotzky (1967) reported laboratory studies of the flocculation of clay suspensions by polymeric products of microbial metabolism. They found that significant flocculation resulted only when the metabolites were added to clays saturated with trivalent or, to a lesser degree, divalent cations. They concluded that organic material was effective only when particle-to-particle repulsion was sufficiently reduced to allow bridging by the metabolites. Pusch (1973a) reported that in inorganic clays and in clays of moderate organic content (<5%), the fabric is similar and controlled by the influence of the electrolyte. Clays in which plant residues predominate show a fabric of particles built up upon the interwoven plant fibers. When long polymeric organic molecules that are capable of joining particles by bridging are present in suspensions undergoing shear, particles of unequal size are joined preferentially (heterocoagulation), according to van der Ven (1981), because the bridging polymer is most easily ruptured by hydrodynamic forces when the particles are nearly equal in size. This preferential heterocoagulation is a sensitive function of the polymer molecular size and the rate of shear, because in the absence of polymer bridging, homocoagulation is generally favored at all rates of shear.

Ingles (1968) reviewed the types of chemical bonding between clay crystals. He stated that edge-to-edge flocculation is the state of least potential energy, even though the ragged nature of the domain edges reduces the area of contact. Face-to-face flocculation is similar essentially to the bonding within the domain. Although energetically favored, edge-to-edge flocculation may not predominate: The limited contact area results in a weak, easily reoriented mass, and the tendency of small particles to adhere to the domain edges may block them before an edge-to-edge structure can be formed. Gillott (1969) speculated that fabric, initially determined by the salinity of the deposition medium, may be preserved by the precipitation of carbonate cements at the points of particle contact and may control the geometry of subsequent carbonate precipitation in the pores.

Leonards and Altschaeffl (1964) and Kassiff and Baker (1971) concluded that although clay particles are not in contact in natural sediments, they are so close to each other that water molecules at points of near-contact reorient and strengthen the sediment. This reorientation was reported to require time periods on the order of days. Mitchell et al. (1969) applied the theory of absolute reaction rates to the time-dependent deformation of soils. They found that the energy required to initiate creep is of the same order of magnitude for wet clay, dry clay, and dry sand, and is much greater than that for the flow of water. They speculated that although the initial fabric is determined by depositional environment, subsequent changes in the fabric under stress are controlled by short-range forces at solid-to-solid interparticle contacts. Andersland and Douglas (1970) reported that the energy needed to initiate creep when nonpolar carbon tetrachloride is the pore fluid is similar to the values found by Mitchell et al. (1969) when water is the pore fluid. Marshall and Kitchener (1966) concluded that the rate of deposition of particles on surfaces is a function of the ionic strength of the suspension medium. Ruckenstein and Prieve (1976) reported that particle deposition depends upon the chemical constitution of the particles and the suspension medium through the influence of changes in the double layer on the potential energy of interaction. They reported that deposition also depends on the size and the shape of the particle through their effects on the dynamics of particle movement. They stated that if reversibility is considered, floc size could be calculated in principle.

Hogg and Yang (1976) performed a theoretical investigation of the effects of shallow secondary minima in potential energy curves (see fig. 2.14 and curve 5, fig. 2.15 and discussion thereof) on aggregation. They concluded that aggregation at the secondary minimum becomes more important than that at the primary minimum as the concentration of ions in the suspension medium decreases and as the particle size increases. For particles of 1-μm size, aggregation at the secondary minimum is predicted to be important at univalent electrolyte concentrations as small as 1 micromole per liter, even though consideration of the primary minimum alone suggests that the electrolyte concentration would have to be at least 1000 times greater than this for aggregation to become significant.

Although studies of the fundamentals of particle interaction such as these have been extremely helpful in developing an understanding of clay physico-chemistry, it is important to emphasize that the potential energy curves have been computed for radically simplified and idealized systems. Advances in such computations are being made, as shown, for example, by publication of the calculated interactions between a semi-infinite plate and an ellipsoidal particle (Brenner and Gajdos 1981), between two cylinders (James and Williams 1981), and between a pair of spherical colloidal particles (Feke et al. 1984). Recently, research groups have measured the forces between two molecularly smooth mica surfaces at close approach (Israelachvili and Pashley 1983; Klein 1983; Pashley and Quirk 1984; Christenson 1984). Much remains to be done before realistic simulations of natural sediments can be made.

Response of Specific Clay Minerals

Studies of the characteristic responses of different clay minerals to exposure to negatively charged ions (Whitehouse and Jeffrey 1955) and to long-term exposure to seawater (Whitehouse and McCarter 1958) have been published. They show that not only are the responses of kaolinite, montmorillonite, and chlorite different from each other at initial contact, but also the minerals themselves were altered slowly (over 5 years or more). Powers (1957, 1959) reported that diluted seawater (ranging from 1/3 to 1/30 the salinity of seawater) is effective in flocculating all the important clay types when clays are in water. Whitehouse et al. (1960) reported that clay minerals settle from brackish waters and seawater differentially or not at all, with the differentiation based upon the characteristic mineral composition of the clay. Illitic materials settle more rapidly than kaolinitic materials, which settle much more rapidly than montmorillonitic materials (smectites). This order of flocculation and settling is independent of the electrolyte composition and concentration and of the presence of a wide variety of organic materials. Only smectites possess settling velocities that depend on salt concentration once the water becomes slightly salty (0.4% by weight), and the velocities for these increase exponentially with salt concentration up to typical seawater concentrations (3.6%). Edzwald and O'Melra (1975) concluded that in brackish water, illitic clay settles more slowly than montmorillionitic and kaolinitic clays when shearing forces in the suspension are kept small.

From studies of kaolinite, Schofield and Sampson (1954) deduced that although the faces of the particles always bear a negative charge, the edges may become positive under some chemical conditions. By changing the composition of a kaolinite suspension, they obtained three fabrics: Deflocculated kaolinite, flocculated kaolinite with particles arranged predominantly face-to-face, or flocculated kaolinite with particles edge-to-face. In their determinations, they included measurements of the apparent viscosity and the volume of sediment produced under various conditions. These results have been of great interest and have been extended and confirmed by Rand and Melton (1977; Melton and Rand 1977). Michaels and Bolger (1962) showed that flocculated kaolinite suspensions under shear contain aggregated flocs as the flow unit and that, as the shear increases, the size of the aggregates decreases. They indicated that face-to-face aggregates are generally smaller than edge-to-face aggregates. Williams and Williams (1978) used measurements of the mobility of kaolinite particles in electric fields to obtain the zeta potential. In agreement with Schofield and

Sampson (1954), they concluded that the faces of the particles are negative, and the edges of the particles are negative in basic solutions, but they are positive in acidic solution. Welzen et al. (1981) measured the influence of surface-active agents on a variety of properties of kaolinite suspensions and concluded that these effects could be explained satisfactorily by application of double-layer theory provided that appropriate edge and face charges are assumed. Lockhart (1980a) concluded from studies of the electrical properties of kaolinite that packets made up of a few (5–7) particles lying face-to-face were aggregated edge-to-face and edge-to-edge, making microscopic structures.

O'Brien (1971) studied kaolinite and illite floccules deposited from distilled water and dilute NaCl solution. He concluded that the double layer was compressed, and this allowed the particles to approach each other and rotate to a parallel or subparallel positon. Thus the resultant fabric consisted largely of stepped face-to-face units. From the consolidation properties of smectites, Klausner and Shainberg (1971) concluded that packets of particles formed extensively in calcium-saturated clays. They stated further that in the presence of both cations, sodium ions tended to be situated outside the packets, whereas calcium ions tended to occupy positions within the packets. Sides and Barden (1971) investigated the fabric of several clays deposited from a variety of media. Chemical additives caused little visible effect on the structure of kaolinitic clays, but their presence was apparent in illitic clays. Sides and Barden concluded that chemical additives become effective in producing dispersed and flocculated structure only as the average clay particle size is reduced and the colloidal activity increases. Lockhart (1980b) concluded from electrical studies of montmorillonite and the synthetic hectorite clay laponite that both form edge-face and edge-edge aggregates in distilled water as the clay concentration increases.

Olson and Mesri (1970) studied the compression and swelling of a variety of minerals. They concluded that mechanical effects predominate for mica, kaolinite, and sands; physico-chemical effects predominate for smectites, and illite exhibits intermediate behavior. Sridharan and Rao (1971) concluded that physico-chemical effects predominate for illite, too. Of course, dealing with essentially monomineralic clays is an idealization if, as was concluded by Weaver and Pollard (1973), most clay particles are interstratified layers of differing composition. Mitchell et al. (1973) reported that electrolytes can spontaneously reduce the volume of very fine-grained soils that are saturated with water by releasing some of the water. Mesri and Olson (1970) reported that for the full variety of fine-grained materials studied, physico-chemical effects are always important in determining the initial fabric of the sediments. Kranck (1980) concluded that fine-grained particles, regardless of mineralogy, are flocculated in most natural waters before reaching the sediment-water interface. Cowking et al. (1983) studied a saponite clay, which occurs as a vein filling in basalt. They concluded that a fibrous form of the clay consists mainly of strongly crystalline laths, and a granular form consists of turbostratically ordered particles. Both forms swell in water and in ethylene glycol. With glycerol, the granular form swells, whereas the fibrous form swells only when it was subjected to vigorous chemical pretreatment or mechanical grinding. Cowking et al. (1983) concluded that this difference in behavior reflects the much more greatly developed order in the fibrous form, which allows its particles to participate more fully in layer-to-layer bonding.

3 Concepts of Clay Fabric

Early Concepts

Perhaps the earliest concepts of clay fabric from engineering and geological perspectives were presented by Terzaghi (1925) in a discussion of the arrangement of soil particles and the bonding of cohesive soils. As early as 1908, Sorby alluded to the importance of clay particle arrangement during settling and consolidation. Terzaghi proposed that clay minerals stick to each other at points of contact, with forces sufficiently strong to construct a honeycomb structure, which permits large amounts of water to be enclosed within the voids. Each unit or cell of the honeycomb was envisioned as being made up of numerous single grains held together by adhesion. Later, Casagrande (1932) presented the hypothesis that the honeycomb structure of sensitive soils is very similar to Terzaghi's concept. This type of clay fabric is referred to as the "*Terzaghi-Casagrande honeycomb structure*" (fig. 3.1). After publication of Terzaghi's work, Goldschmidt (1926) proposed that clay properties are due to crystalline minerals enclosed by a film of adsorbed water molecules. He thought that the water molecules adhere to each other and to mineral grains due to their dipolar moment. Goldschmidt proposed that the flaky minerals of chemically sensitive clays are arranged in an unstable cardhouse structure. This structure was assumed to show a dominance of particle edge-to-face contacts. Over twenty years later, Lambe (1953) presented diagrams of clay mineral arrangements that are similar to Goldschmidt's concept of clay fabric (fig. 3.2). Lambe proposed that undisturbed marine clays possess an open structure similar to the cardhouse structure; however, the fabric of freshwater clays was thought to be somewhat denser (fig. 3.3). In remolded clay, it was assumed that there is a greater degree of parallelism between adjacent particles (fig. 3.4). Clearly, the Goldschmidt-Lambe concept of a cardhouse structure is very different from the Terzaghi-Casagrande honeycomb concept. Rosenqvist (1955) and Bjerrum (Bjerrum, Rosenqvist 1956) presented similar views of the cardhouse fabric; however, their means of optical light microscopy were not completely satisfactory for making definitive conclusions. The cardhouse fabric also was discussed and proposed by Hoffmann as early as 1942 (Hoffmann 1942), although his work was published much later than Goldschmidt's initial studies.

Keller (1936) drew attention to the importance of clay colloids as a cause of bedding in shale. He pointed out also that in contrast to

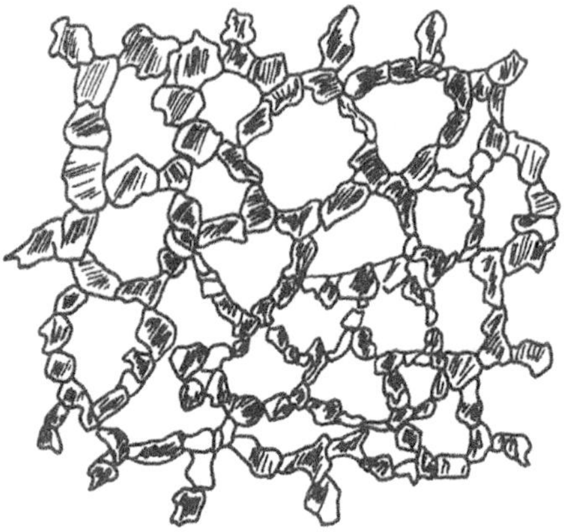

Figure 3.1 *Honeycomb structure (redrawn from Terzaghi 1925 and Casagrande 1932).*

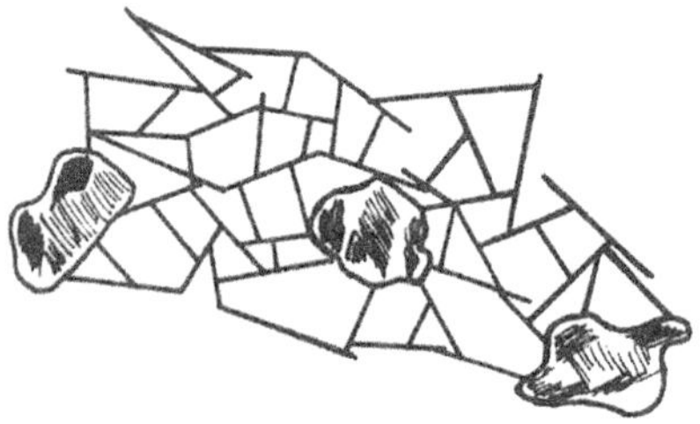

Figure 3.2 *Cardhouse structure of saltwater deposit (redrawn from Goldschmidt 1926 and Lambe 1953).*

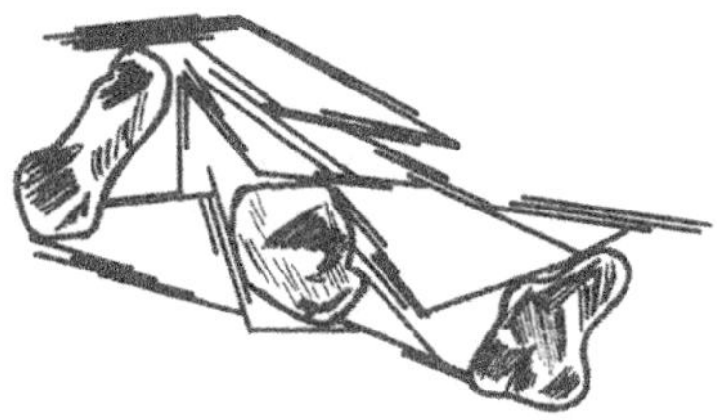

Figure 3.3 *Cardhouse structure of freshwater deposit (redrawn from Lambe 1953).*

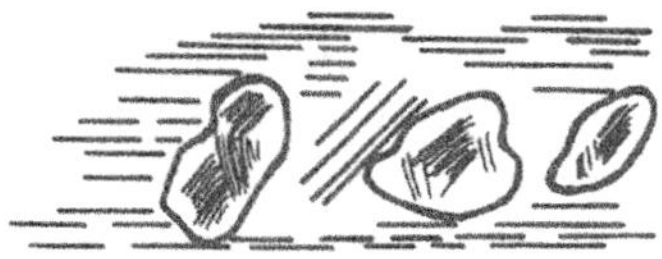

Figure 3.4 *Remolded structure (redrawn from Lambe 1953).*

Figure 3.5 *Three-dimensional representation of clay structure (redrawn from Tan 1957).*

some shales, some massive clay deposits are characterized by random orientation of clay flakes (Keller 1936, 1946). Keller recognized the importance of the chemical environment on flocculation and settling of clay colloids and suggested different modes of origin for shales and nonbedded fireclays. Keller (1936, 57) stated that ". . . colloidal clay brought to a saline sea is not expected to be carried far from shore. Much or most of it will be flocculated upon contact with the seawater because of the high concentration of ions in the latter." He believed that some clay, however, brought to the sea would remain dispersed for a considerable time, because "some one of the great variety of ions available during the history of the clay for adsorption in the complex should be stabilizing." During the nearly 50 years that passed since the appearance of this publication, supporting evidence for the first but not the second conjecture accumulated.

Lambe (1953) thought that the random arrangement (cardhouse fabric) of the particles occurs when clay sediment enters salt water with a large concentration of electrolyte. Flocculation is rapid, because the salt removes the energy barrier that tends to keep particles apart (see fig. 2.15 for the effect of increasing electrolyte concentation on the energy barrier between particles). Rapid flocculation of these particles causes them to stick to one another at the random orientations of initial particle approach and to incorporate into the open flocs silt and fine sand particles from the water column. Clay deposited in fresh water, however, remains dispersed and settles with a slower velocity than does silt, and it forms a much greater degree of parallel particle orientation (oriented fabric). This occurs because the particles must overcome an energy barrier to aggregate, and their potential energy is minimized (maximum van der Waal's attraction) when platey particles approach face to face. Once particles have settled to the sediment-water interface, they tend to reorient to the minimum energy position, that is, the parallel particle orientation. The particles become stabilized in this position by being driven sufficiently close together by the downward force of sediment subsequently deposited over them. Thus according to Lambe (1953), particle orientation in a dispersed system is a parallel arrangement (oriented fabric), whereas in a flocculated system, it is random (cardhouse fabric).

In an effort to reconstruct the environments of deposition of argillaceous rocks, Urbain (1937) used the optical microscope to delineate the microfabrics. He recognized the importance of the effects of electrolytes on the clay particle arrangements. Powers (1957) investigated clay floccules of suspended sediment from various estuaries. His studies indicated a decrease in floccule size and a slight increase in crystal size in the down stream (estuary) direction. In Chesapeake Bay, he found no size sorting of clay minerals in the suspended sediment. In Atchafalaya Bay, his studies showed that at 1–10 parts per thousand salinity, flocculation and rapid deposition of mud occurred.

Mitchell (1956) pointed out three important differences between dispersed and flocculated clays in relation to their geotechnical properties:

1. For any given consolidation pressure, the weight of clay occupies a smaller volume in the dispersed or oriented state than in a flocculated condition.
2. Dispersed clay particles are distributed more uniformly throughout a given volume than are particles in a flocculated clay.
3. For a given increment of stress, the shifting of particles relative to one another is greater in a flocculated clay than in a dispersed clay.

Fabric studies by Mitchell (1956) of undisturbed and remolded clays revealed that silt particles are not in contact with each other in either a remolded or undisturbed state. The silt appeared to float in a clay matrix, even in soils in which more than 75% of the particles were larger than 2 μm. Remolding tends to homogenize the sample, rendering the clay less variable. The improvement in the parallel orientation of clay platelets was greater in remolded samples than in undisturbed samples when they were compressed in one dimension at natural water contents to loads of 196 kPa. Remolding caused preferred orientation over small areas, such that oriented areas shifted smoothly from particle alignment in one direction to particle alignment in another direction. Over large areas, preferred particle orientation parallel to one direction was not strong; apparently, kneading is responsible for this, because there is no force that tends to orient particles in one direction. The changes in grain orientation and engineering properties as a result of remolding were greater for the illite-chlorite group of marine clays than for illite-montmorillonite freshwater clays and Mexico

City clay, which is a weathered volcanic glass. Mitchell (1956) found also that the clays deposited in marine or brackish water that had undergone natural precompression of up to 284 kPa (1 psi = 6.89 kPa) exhibited some degree of parallel particle orientation. Rosenqvist (1955) reported that chemically sensitive Norwegian marine clays showed almost complete random particle orientation, although they had never been consolidated to a great degree. Mitchell concluded that the concepts of Lambe (1953) were essentially correct for freshwater, marine, and remolded clays. However, deposition of clay in fresh water may not, in every case, produce an oriented fabric, because impurities in the depositional water and the nature of the adsorbed cations may affect the dispersion of the clay platelets during sedimentation. Readers are referred to early, detailed studies of compaction, engineering properties, and clay fabric by Seed and Chan (1959), Trollope and Chan (1960), and Lambe (1960). Mechanical aspects of clay fabric have been studied also by the ceramicists Weymouth and Williamson (1953), Williams and Williams (1947, 1960a, 1960b).

Early concepts of clay fabric were depicted in two-dimensional figures that represent the three-dimensional arrangements of clay particles. Later, Tan (1957, 1959) presented a schematic of clay fabric in three dimensions (fig. 3.5), which showed a clay mineral network dominated by contacts between corners and planes of mineral particles. Using electron microscopy (carbon replication), Rosenqvist (1959) confirmed the Goldschmidt-Lambe cardhouse concept for undisturbed clay. The mineral arrangement was dominated by contacts between corners and planes, and it was similar to the fabric proposed by Tan. This arrangement was found for all marine clays examined by Rosenqvist. Complete random orientation of illite and kaolinite particles in marine clay from the northeastern area of the Black Sea was reported by Raitburd (1960). Different degrees of random and preferred orientation was reported by Wu (1958) for glacial-like clays from the Great Lakes region. The arrangements could not be related, however, to the types and concentrations of ions.

Lambe (1958a) described concepts of clay fabric that differed slightly from those he had proposed during 1953 (figs. 3.6, 3.7, and 3.8). He depicted salt-type flocculation with grain orientation approaching parallelism and a somewhat open network indicative of a moderately high void ratio (fig. 3.6). The non-salt-type fabric exhibited a purely edge-to-face relationship among grains, resulting in a large void ratio framework (fig. 3.7). The dispersed particle arrangement (fig. 3.8) is quite similar to the representation he presented earlier. An essential difference between the fabric concepts is the absence of silt particles in the more recent representation. Similar clay mineral arrangements were shown by Schofield and Samson (1954). Other studies indicate that in sands permeated by clay suspensions, the clay plates orient parallel to the sand surfaces (Brewer and Haldane 1957). Clearly, the presence of silt and sand mixed with clays would influence the overall arrangements of the clay particles in close proximity to the larger grains.

Studies by Lambe (1958a) with pure kaolinite indicated that the particles flocculate with an edge-to-face arrangement in pure water, and a small amount of salt results in dispersion. (Even in suspensions of pure kaolinite, there would be mobile, positive ions from the clay in the water phase, small concentrations of hydrogen and hydroxide ions due to autoprotolysis of water, and perhaps ions from dissolved gases, such as carbon dioxide. In solutions of minimal ionic strength, the fabric may be sensitive also to small changes in hydrodynamic conditions.) Large concentrations of salt result in flocculation, but the increase in anions that gather at the particle edges results in a less effective positive charge over a smaller distance, yielding a smaller degree of flocculation (fig. 3.6). Readers are referred to a companion paper by Lambe (1958b), in which the relations between fabric and the engineering behavior of clay are discussed.

van Olphen (1963) depicted in a descriptive sense the difference between the particle arrangements due to sedimentation from peptized versus flocculated suspensions (fig. 3.9). The main intent was to show that flocculated particles form a loose, voluminous sediment, whereas a relatively dense sediment of closely packed particles forms from a peptized suspension. This could be equivalent to Lambe's flocculated versus dispersed fabrics, respectively. Models proposed by von Engelhardt and Gaida (1963) depict voluminous clay structures (large aggregates), which are indicative of high-salinity water and a denser, small-aggregate structure characteristic of lower-salinity water. van Olphen (1977) also described various conceptual models of possible modes of particle association (fig. 3.10). In his scheme, associations built up of single or "dispersed" particles are contrasted with associations composed of compound or "aggregated" particles (fig. 3.10). van Olphen pointed out that three modes of particle association are possible

Figure 3.6 *Salt-flocculated card-house structure with a stairstep configuration (redrawn from Lambe 1958).*

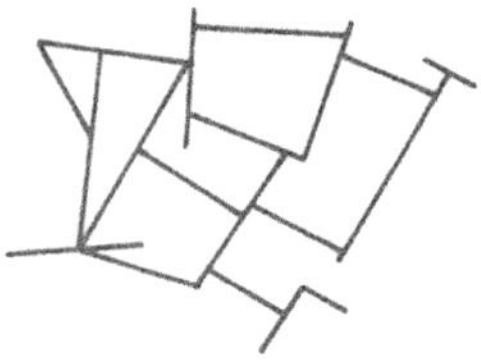

Figure 3.7 *Non-salt-flocculated, well-developed cardhouse structure (redrawn from Lambe 1958).*

Figure 3.8 *Dispersed structure with preferred grain orientation (redrawn from Lambe 1958).*

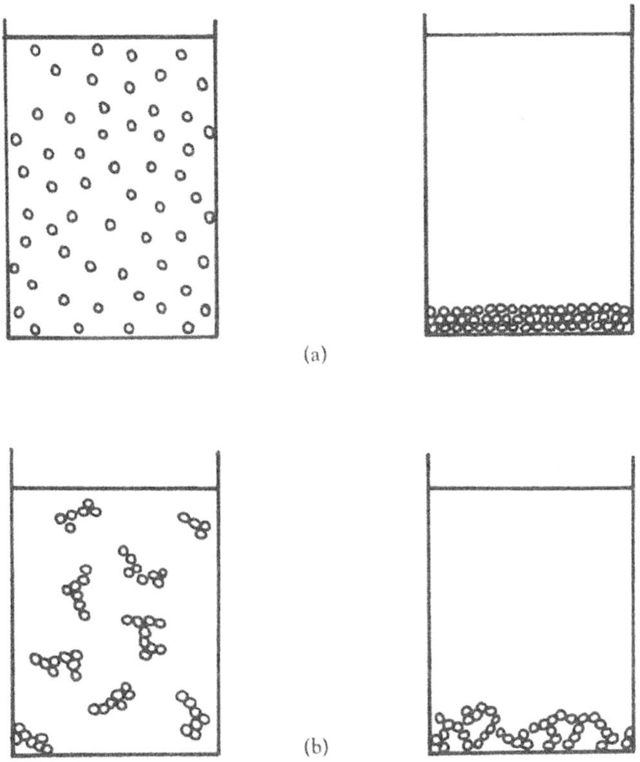

Figure 3.9 *(a) Sedimentation from a peptized suspension results in a dense, close-packed sediment. (b) Sedimentation of flocculated suspension produces a loose, voluminous sediment (after van Olphen 1963).*

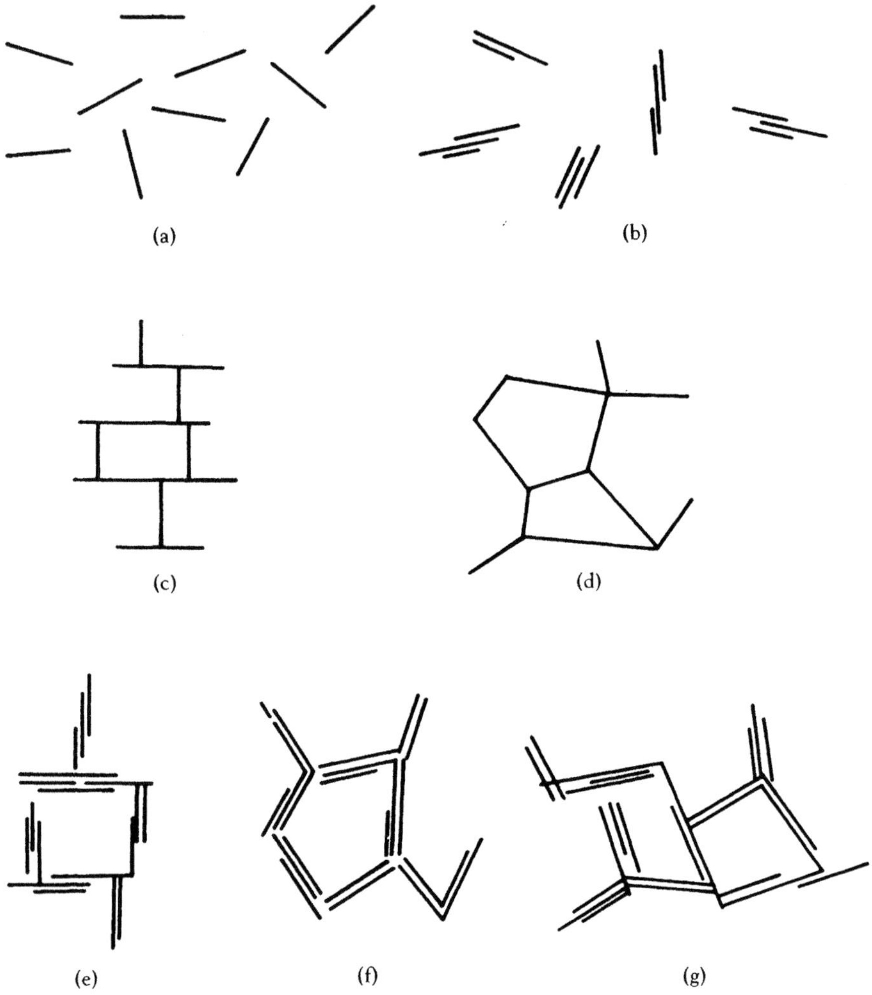

Figure 3.10 *Various modes of particle association (after van Olphen 1963): (a) Dispersed and deflocculated. (b) Aggregated but deflocculated, (c) EF flocculated but dispersed, (d) EE flocculated but dispersed, (e) EF and FF flocculated and aggregated, (f) EE flocculated and aggregated, (g) EF and EE flocculated and aggregated.*

when platelike clay particles flocculate: Face-to-face (FF), edge-to-face (EF), and edge-to-edge (EE). Only the EE and EF types of particle association produce agglomerates, which are called "flocs." The FF association is termed "aggregation" and refers not only to several FF multi-layer particles but also to the degree of unit-layer stacking in a single particle. The type of particle association so produced depends strongly on the physico-chemistry of the suspension during sedimentation. Apparently, all the structures described by van Olphen have not been found for natural sediments (Moon 1972). The models, however, do afford a means of reference for the study of fabric in natural clay-water systems.

Prior to 1962, a few fabric studies of naturally consolidated clayey sediments had been published (Grim et al. 1957, Kaarsberg 1959, White 1961, Meade 1961). In an excellent review of particle arrangement during compaction (consolidation) of clayey sediment, Meade (1964) contended that there was meager evidence showing that preferred and turbostratic fabrics (Aylmore and Quirk 1960) are formed readily and generally during natural compaction (consolidation), although these fabrics could be reproduced in laboratory tests.

Using X-ray diffraction, Quigley and Thompson (1966) found that reorientation of clay platelets occurred during anisotropic consolidation of Leda marine clay, and an abrupt increase in parallelism occurred at pressures exceeding the preconsolidation pressure. In undisturbed samples, reorientation occurred in a plane perpendicular to the direction of the major principal consolidation pressure. At lower pressures, the resulting particle parallelism was greater in remolded samples than in undisturbed clays, but the relation between fabric and void ratio was the same for both undisturbed and remolded samples. Their observation of greater preferred particle orientation for remolded soils agrees with results reported earlier by Mitchell (1956). Using artificially prepared kaolinite samples (mixed with distilled water) and an electron microscope, Smart (1967b) found that the unconsolidated samples exhibited random arrangements of particles or of small groups of particles. Failure zones in sheared samples showed preferred orientation of particles parallel to the zone of shear. Preferred orientation and turbostratic structure developed during consolidation testing. Smart also reported that nothing similar to a cardhouse structure or salt-flocculated fabric was observed in any of the samples tested (see also work by Morgenstern and Tchalenko [1967] on structures observed in kaolin subjected to direct shear).

Later Concepts and Observations

During the past 15 to 20 years, studies of clay fabric have gained momentum with the advent of the electron microscope (EM), and renewed interest has resulted in numerous investigations of clay particle arrangements in sediment by direct observation. Early studies by Rosenqvist (1959) set the stage for EM studies of clay fabric conducted during the 1960s. The first investigations of clay minerals using the electron microscope were carried out by Eitel et al. (1939, Eitel and Schusterius 1940), but they were concerned primarily with the sizes and shapes of the very small particles rather than the depositional fabric. Hast (1947) also employed the transmission electron microscope in the study of clays. Although Hast concentrated on techniques, he noted that clay consists of small planar crystals, such as rounded plates, and these plates are placed one upon the other, with a slight displacement of each.

Aylmore and Quirk (1960) proposed the term *turbostratic arrangement* for a fabric consisting of domains or stacks (Aylmore and Quirk 1959; Sides and Barden 1971); they are referred to as *tactoids*, *packets*, or *clusters* of clay platelets, among which the preferred orientation is nearly perfect (figs. 3.11 and 3.12). Kell (1964) and Sloane and Kell (1966) described a clay fabric that is similar to the turbostratic arrangement, but it has more open structure, which they call *"bookhouse" fabric* or *book structure* (fig. 3.13). Ingles (1968) referred to this fabric as *card-house*. Evidence supporting this turbostratic-type fabric was given by Rosenqvist (1959), although the term had not been proposed by that time. Types of domains were described by Sloane and Kell (1966) as *books* (fig. 3.14(*a*)) and by Smalley and Cabrera (1969) as *stepped face-to-face* (fig. 3.14(*b*), (*c*)). In retrospect, it appears that Hast (1947) implied the stepped face-to-face type of domain, and the book fabric seems to be equivalent to the aggregated particle described by van Olphen (1963). Confusion surrounding terms used to describe regions of near-parallel alignment of plate-shaped crystals in clay minerals led Quirk and Aylmore (1971) to review the usage of the terms tactoid, crystal of montmorillonite, quasi-crystal, and domain. These authors recommended using the term *domain* to describe the regions of parallel alignment of crystals of illite and other fixed-

Figure 3.11 *Turbostratic structure (redrawn from Aylmore and Quirk 1960).*

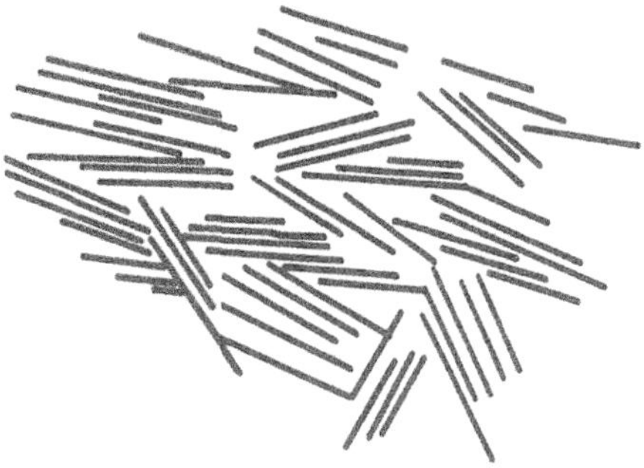

Figure 3.12 *Perfect stack (redrawn from Sides and Barden 1971).*

Figure 3.13 *Bookhouse or book structure (after Kell 1964; modified from Sloan and Kell 1966).*

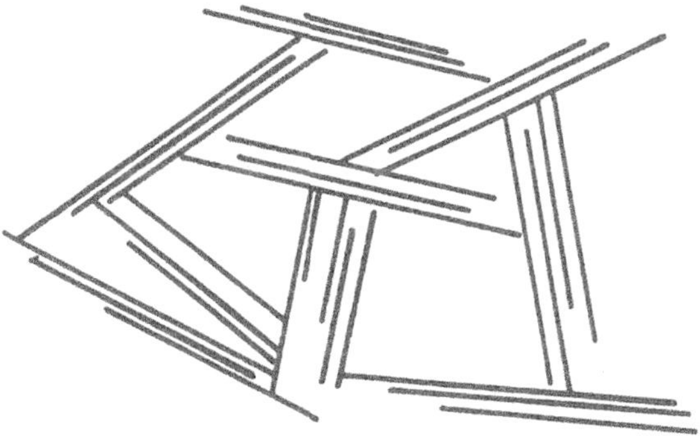

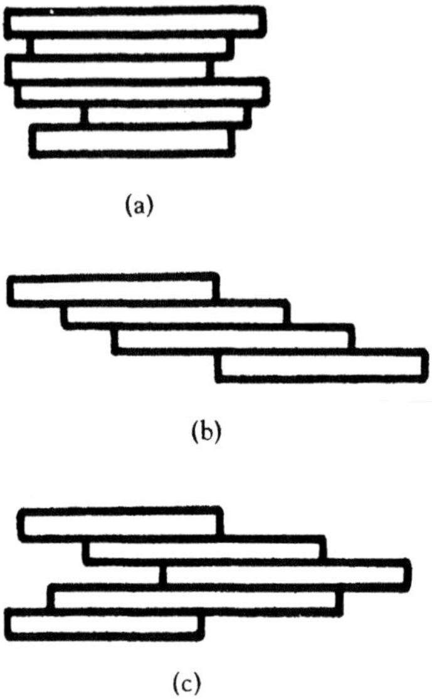

Figure 3.14 *Domain structures (redrawn from Moon 1972): (a) Book (redrawn from Sloan and Kell 1966) and (b, c) Stepped face-to-face structures (redrawn from Smalley and Cabrera 1969).*

lattice clays. The term *quasi-crystals* describes the regions of parallel alignment of individual alumino-silicate lamellae in montmorillonite. Quirk and Aylmore pointed out, however, that the montmorillonite crystal is actually a particular type of domain formed by the alignment of individual alumino-silicate layers. During recent years, the term domain has been used to describe arrays and packings of clay particles in face-to-face arrangements (see Aylmore and Quirk 1959; Moon 1972; Pusch 1973b).

O'Brien and Harrison (1967) investigated clay flake orientation in a sensitive illitic Pleistocene marine clay. They made carbon replicas of air-dried samples and examined them by transmission electron microscopy. Random orientation of clay flakes was the most apparent feature, and the silt grains appeared to float in the clay matrix (these conclusions and observations are similar to those of Mitchell [1956]). They concluded that the random orientation of the clay flakes probably was caused by flocculation in a saline environment. The fabric of chemically sensitive clays was studied also by Karlsson and Pusch (1967, Pusch 1966a). Pusch (1967) carried out extensive, detailed studies and assessed selected clay fabric techniques, including sample preparations and transmission electron microscope (TEM) observations of ultrathin sections. He concluded that sample embedding techniques by diffusion of selected miscible fluids does not affect adversely the original fabric, and artifacts sometimes caused by microtoming are easily delineated. Pusch (1967) investigated the fabric of a very stiff marine clay, which was formed during late Silurian time and had been consolidated by sediments and ice belts. He used a high-voltage microscope that permitted the investigation of water-saturated specimens. He found the fabric was characterized by flocculation similar to soft marine clays. Pusch described the fabric as a system of aggregates linked by chains of small particles. He reported that the water-saturated specimens and specimens treated with acrylate plastic were very similar.

The fabrics of Ordovician shale and mudstone from Ontario were compared to that of post-Pleistocene marine clay (Gillott 1969) by using the petrographic microscope and scanning electron microscope (SEM). Petrographic microscope examination of the fissile shale indicated a strong but not perfect degree of preferred orientation of the micaceous particles. Observation of the mudstone showed little to no preferred orientation among platey minerals. At small SEM magnification (~300x) of the shale, preferred orientation was apparent; however, at higher magnification (~3600x), preferred grain orientation was less apparent, and platey minerals were arranged at large angles to the fissility of the shale. Apparently, fissility in some shales does not require perfectly oriented particles, but it depends on a sufficiently large number of grains with parallel arrangement to establish the fissile property. Observations of the mudstone with the SEM confirmed the light microscope observations of random particle orientation. Scanning electron micrographs of Leda clay revealed random particle orientation in some samples and preferred orientation in others. The Leda clay samples had been subjected to a natural preconsolidation pressure of about 490 kPa. Some degree of random arrangement of particles would be expected in these clays deposited under marine conditions. Some photographs, however, showed that the orientation was far from perfect, with many crystal faces and edges at oblique angles.

In a study of quick clays of eastern Canada (specifically, Leda clay), Crawford (1968) showed that little or no particle reorientation occurred at pressures up to the preconsolidation pressure during odometer measurements. He observed an abrupt increase in particle parallelism at stresses greater than the preconsolidation pressure; this observation was reported previously by Quigley and Thompson (1966). Remolded samples revealed greater parallelism of particles and lower void ratios at all pressures. Scanning electron micrographs of the undisturbed samples (freeze-dried and fractured) revealed random orientation of particles in samples that had a preconsolidation pressure (P_c) of 191.5 kPa, whereas samples that had a P_c of 478.8 kPa showed a slight degree of parallelism.

Strong preferred orientation of clay particles in argillaceous rocks has been reported by several investigators. White (1961) reported that random orientation of clay minerals in shales resulted in poor fissility, whereas fissile shales showed clay minerals oriented with the longest dimensions parallel to the bedding. White attributed flocculation in saline environments to random orientation and concluded that in a sedimentary environment of minimal salt concentration, the particles would settle as individual particles with parallel orientation, ultimately producing a well-developed fissile shale (see the discussion of the effects of electrolytes in the physico-chemistry section of this review). Gipson

(1965, 1966), Odom (1967), and O'Brien (1968) discussed the preferred orientation of clay particles associated with fissility and organic material in shales. Martin (1965, 1966) and Martin and Ladd (1970) discussed the determination of clay particle orientation by X-ray measurement.

Electron micrographs of clayey submarine sediment showed that major reduction in void ratio was possible at relatively low pressure during consolidation tests (Bowles et al. 1969). At a pressure of 49 kPa, the fabric exhibited an open but random arrangement of particles. At a pressure of 392 kPa and a void ratio (ratio of the volume of voids to volume of solids) of 1.3, the fabric consisted of a denser packing of clumps and packets of particles with random appearance overall. This would correspond, apparently, to a turbostratic-type fabric. Samples consolidated to 3138 and 6276 kPa revealed a great degree of parallelism among particles, with some randomly arranged particles. In an earlier study of submarine sediments, Bowles (1968b) concluded that the clay particle arrangements of the undisturbed sediment closely resembled Terzaghi's (1925) honeycomb structure.

Photomicrographs of surficial samples recovered from three depositional environments in the Gulf of Mexico demonstrate clearly the presence of substantial porosity created by edge-to-face particle contacts and the presence of particle chains (Bowles 1968, figs. 3.15, 3.16, 3.17). Fabric of sediment from the continental shelf, slope, and abyssal plain is characterized by particles that are predominately randomly arranged. Domains clearly are present, and void sizes are extremely variable. The photomicrographs reveal an important characteristic of the sediment from a two-dimensional perspective: Most voids are discontinuous, which implies that passageways for fluid flow during the natural consolidation process are constricted. For consolidation to proceed, fluid must flow through continuous connected pores. This may occur by breaking particle contacts or flow through passages present in the third dimension of the field of view, or both. If the micrographs are representative of the sample (i.e., random arrangement of particles), then particle contacts probably would break to allow fluid flow to proceed during consolidation.

Ingles (1968) showed schematically how the cardhouse fabric could change (particle reorientation) during consolidation (fig. 3.18). A reduction in the total volume of the voids occurs through

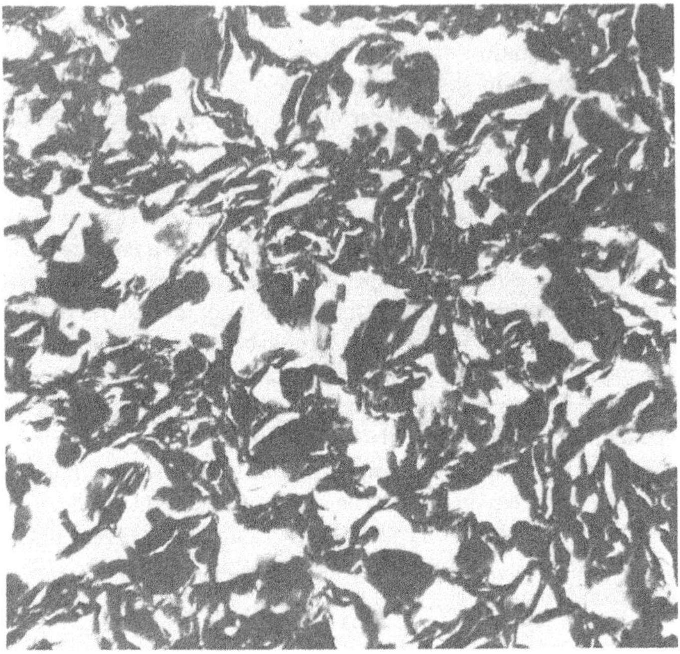

Figure 3.15 *Clay fabric of continental shelf sediment from the Gulf of Mexico. Note the random arrangement of short linking chains and variable sizes of voids. (Magnification 5500X) (Reprinted by permission of Frederick Bowles).*

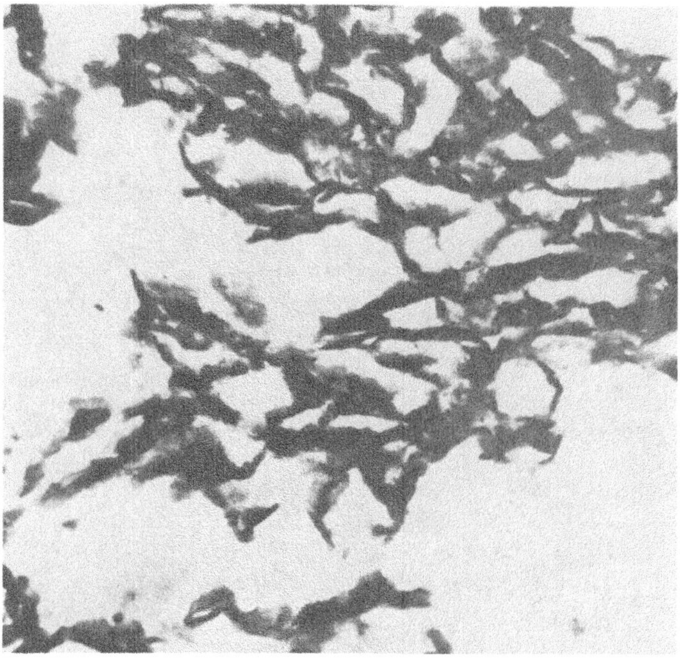

Figure 3.16 *Clay fabric of continental slope sediment from the Gulf of Mexico, showing edge-to-edge and edge-to-face particle contacts, and that voids are large compared to the size of particles. (Magnification 8700X) (Reprinted by permission of Frederick Bowles).*

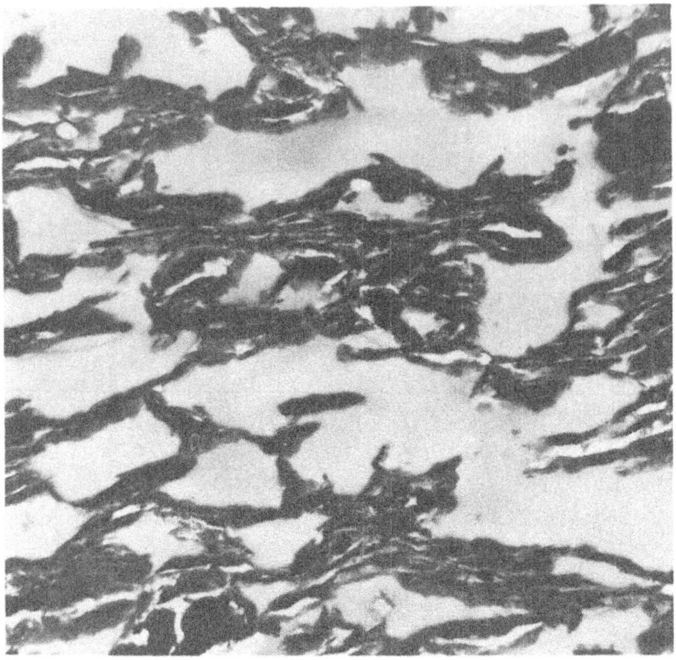

Figure 3.17 *Clay fabric of abyssal plain sediment. Note the short linking chains and the presence of domains. Photomicrograph shows slight tendency toward preferential arrangement of particles. (Magnification 6210X) (Reprinted by permission of Frederick Bowles).*

an increase in the degree of particle orientation. Domains form and increase in size (fig. 3.18(b)) along the virgin consolidation curve and orient wholly in a face-to-face relation under great pressure (fig. 3.18(c)). During consolidation, an edge-to-edge fabric (tactoid model, fig. 3.19) would experience an increase in the area and number of face-to-face bonds and an increase in tactoid links. It is significant that Ingles's cardhouse fabric consists of randomly arranged edge-to-face domains (for an exception, see fig. 3.18(a)) rather than randomly arranged single particles, as was proposed in early concepts of clay fabric (fig. 3.20). This fabric is essentially the same as the bookhouse fabric of Kell (1964) and Sloane and Kell (1966). Ultrathin section photomicrographs of clayey submarine sediment strongly support the concepts proposed by Ingles (1968) for consolidated clay sediment (Bowles et al. 1969; Bennett et al. 1981).

Barden and Sides (1970) investigated the engineering behavior and structure of selected compacted clay samples. Samples of kaolin compacted both wet and dry of optimum (i.e., the water content at which the greatest degree of compaction is attained) revealed no marked difference in structure but showed turbostratic structure at high SEM magnifications. Visual assessment revealed a homogenous structure wet of optimum but pelletlike macropeds dry of optimum. Similar conclusions about compacted structure were made by Sloane and Kell (1966) and Smart (1967b). Tovey (1970) constructively criticized the work of Barden and Sides (1970) and stated that particle associations in undisturbed kaolinite showed stacks of particles in face-to-face arrangements

Cabrera and Smalley (1971) pointed out that the characteristic domain structure or turbostratic structure formed in compacted kaolinite soils, which they described as *stepped face-to-face*, is different from the structure of relatively undisturbed kaolinite systems. Cabrera and Smalley concluded that bookhouse stacks (large particles) are characteristic of undisturbed kaolinite deposits, and during the compaction of kaolinite soil from an undisturbed condition, several events occur:

1. The bookhouse structure is disturbed, with the large kaolinite particles being broken successively, ultimately reaching the true particle size of 1 nm. The smaller the particle size, the less effectively the bulk mechanical forces can be brought to bear on the particles in order to reduce their size further.

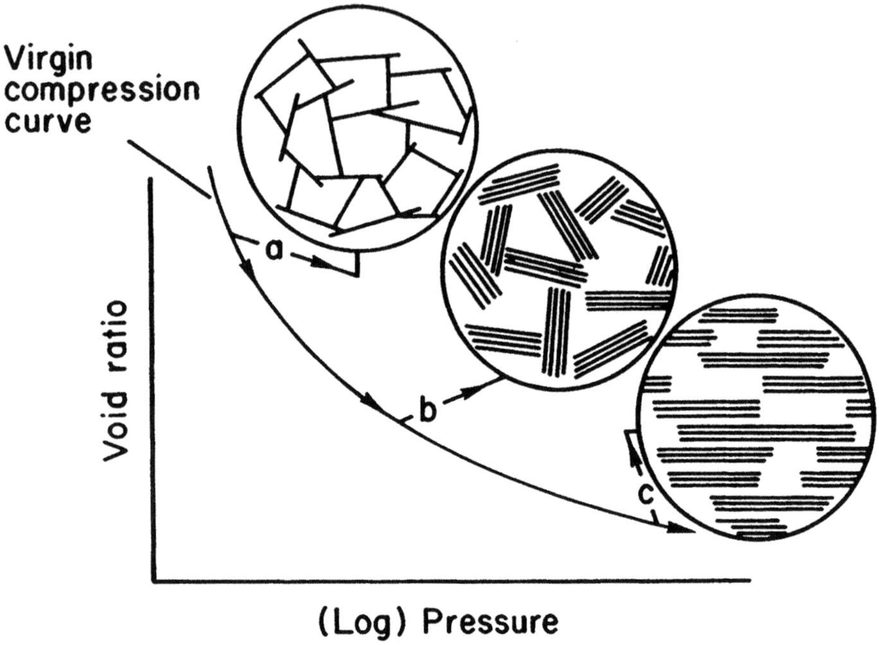

Figure 3.18 *Fabric changes during consolidation in a clay (redrawn from Ingles 1968).*

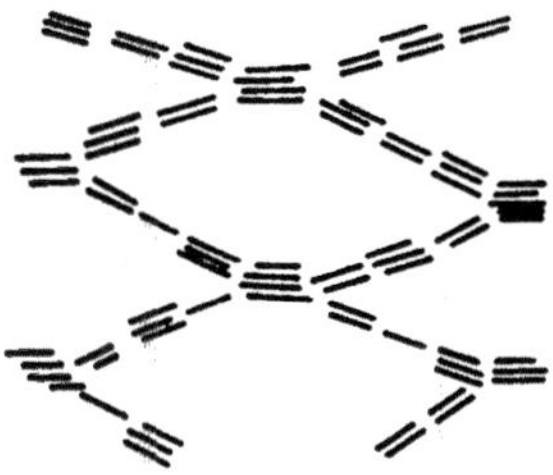

Figure 3.19 *Tactoid structure (redrawn from Ingles 1968).*

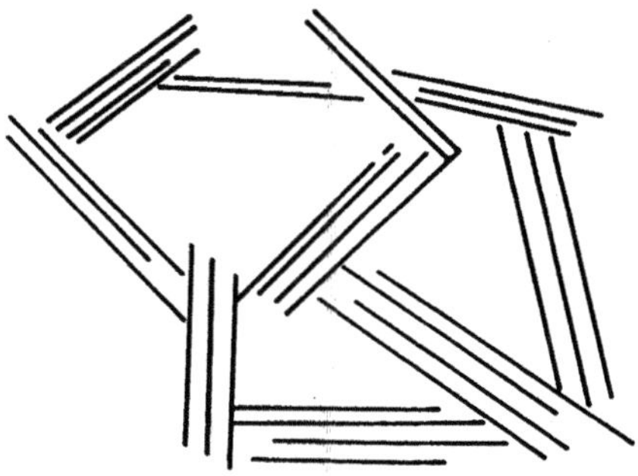

Figure 3.20 *Cardhouse structure (redrawn from Ingles 1968).*

2. Discrete domains reform, but the initial regularity is not restored, and stepped structures form.
3. Stepped structure is predominant through compaction processes, and this structure becomes the form of particle association. The stepped structure is considered to be characteristic of compacted kaolinite soils, and the bookhouse stacks are characteristic of undisturbed kaolinite deposits.

Current Concepts and Observations

Interest in clay microstructure and particularly clay fabric has intensified since about 1970. The acceleration of research probably has been due to not only the numerous findings and observations made by using the electron microscope, but also the realization of the importance of the influence of clay microstructure on the engineering behavior, sedimentological properties, and the fundamental nature of clayey sediment. Meetings and symposia have been devoted almost entirely to the presentation of current research on clay microstructure. There have been conferences in which microstructure was considered in engineering and clay mineralogy studies. Excellent studies were presented at the Southeastern Asian conferences on soil engineering (particularly during 1970 and 1971), the Roscoe Memorial Symposium (Cambridge University , 1971), the Third International Conference on Expansive Soils (Haifa, Israel 1973), the Fourth International Working-Meeting on Soil Micromorphology (Kingston, Ontario, Canada, 1973), the Clays and Clay Minerals Conference (Banff, Canada, 1973), and the International Symposium on Soil Structure, (Gothenburg, Sweden, 1973). In the earliest studies using the electron microscope, TEM was employed exclusively. Recently, however, SEM has been employed frequently in studying the fractured surfaces of clay samples.

SEM studies of laboratory-sedimented kaolinite and illite revealed little difference between the gross structures of kaolinite and illite floccules formed in distilled water (large clay concentration) and floccules formed in salt water (small clay concentration) with 1 g/l of NaCl (O'Brien 1970b, 1971). O'Brien suggested that under experimental conditions, the double layer is compressed, permitting the clay flakes to approach one another closely, resulting in significant van der Waals forces of attraction (see the discussion on the responses of specific clay minerals in chap. 2). Flat flakes of kaolinite would approach each other in a parallel position, and the warped illite flakes would arrange in a subparallel position. The resulting structures would be oriented arrangements of stepped clusters (figs. 3.21 and 3.22). O'Brien considered the honeycomb concept of Terzaghi (1925a) to be more representative of the observed clay fabric than was the cardhouse model of Casagrande (1940) and Tan (1957, 1959), but the fabric was more complex than depicted by the early diagrams. Uncompressed flocculated kaolinite and illite were considered to consist of a very porous network of randomly oriented flakes or clumps. A single floccule unit of kaolinite showed numerous face-to-face flakes in a cluster with a stairstep arrangement. The domains may be oriented at any angle and attached in edge-to-face arrangements. This representation appears to be quite similar to Lambe's (1958a) concept of salt-flocculated clays. Smalley and Cabrera (1969) also described a similar fabric that they called "stepped face-to-face." O'Brien's illite floccule unit consists of several stepped face-to-face oriented flakes.

A more recent experimental study by Lanier and Jones (1979) of the textural relationships of kaolinite clay flocculated in increasingly saline solutions and slightly consolidated, showed apparently significantly different particle associations compared with studies by O'Brien (1970c, 1971). Using SEM and observations during settling of clay suspensions, several conclusions were made:

1. Settling rates were slowest in solutions with the largest salinities.
2. The greater the kaolinite concentration, the faster the rate at which flocculation occurred.
3. Kaolinite suspensions flocculated in distilled and slightly saline water (1% NaCl) were characterized by individual grains linked in face-to-face contact in an offset *(en echelon)* pattern that formed simple grain aggregates (chains and small platelike units). After some consolidation, this fabric was preserved only partially.
4. Kaolinte suspensions flocculated in solutions with intermediate salinities (1–3%) showed chains that increased in size and frequency. These chains cross-linked to form broad platelike aggregates. This particle association seems to be similar to the "shingle-type floc" of Mathewson et al. (1973). The platelike units joined in edge-to-face linkage with other units and formed a framework that was less susceptible to collapse.

Figure 3.21 *Stairstep structure of kaolinite minerals (redrawn from O'Brien 1971).*

Figure 3.22 *Stairstep structure of illite minerals (redrawn from O'Brien 1971).*

5. An apparent increase in the frequency of preservation of the edge-to-face linked platelike aggregates was observed in suspensions flocculated in solutions with salinities that were approximately that of seawater (3.6% NaCl). This type of fabric resulted in an open internal floccule framework with very high porosity.

Lanier and Jones (1979) concluded that the samples flocculated in intermediate and normal salinities appeared to possess the greatest internal strength in comparison with samples flocculated at lower salinities. Kranck (1976) found that the textures of flocculated marine sediments of ancient silty and muddy rocks display some features that are similar to textures formed by flocculation of suspended sediment.

During SEM investigations of suspended sediments from the Brazos River, Texas, Mathewson et al. (personal communication 1973) observed an increase in floccule size in the downriver direction, which corresponds to an increase in salinity. Their observation conflicted with observations made by Powers (1957) who studied suspended sediments in estuaries. The apparent conflict may be in defining floccule versus crystal size. Clay particles were arranged predominantly in face-to-face contact, forming a shingle-type floc. Also, surficial submarine sediments were characterized by a shingle-house fabric of high porosity. The shingle became relatively large upon contact with water characterized by salinities of approximately 1.5%, and although the large shingle is in suspension, it behaved as a flexible sheet (similar to a two-dimensional form). According to Mathewson, when a large sheet experiences turbulence in the coastal environment, it is rolled and becomes a three-dimensional, thin-shelled floccule. Mathewson (personal communication 1973) also hypothesizes that the shingle-house sedimentary fabric unit behaves, in an engineering sense, as a rigid, thin-shelled structure. He suggested that this fabric would explain the apparent overconsolidation of numerous surficial submarine sediments, and would be compatible with the engineering requirements discussed by Lambe (1953).

A study of shale and claystone fabric established a definite correlation between clay-flake orientation and fissility (O'Brien 1970b). Certain organic-rich black shales revealed the best fissility and greatest degree of preferred grain orientation. Shales with poor to moderate fissility showed gross parallelism of flakes with areas of platelets dipping at angles to the bedding. Random clay flake orientation was observed only in nonfissile claystones.

O'Brien (1970b) proposed three possible means of producing parallel orientation of clay flakes in argillaceous rocks: (1) Original deposition in parallel arrangements, (2) mechanical rearrangement and orientation during the early stages of compaction, and (3) preferential growth of layer silicates under consolidation.

Geochemical factors affecting parallel orientation of clay grains could be caused by the presence of organic molecules that neutralize surface charges of the clays and result in sedimentation in a dispersed state (O'Brien 1970b). In addition, clay concentrations and small rates of deposition may result in sedimentation in a dispersed state even in the presence of high salinities (Rosenqvist 1966). Thus well-developed fissility in shales of marine origin indicates dilute clay concentrations in sea water. Shales with poor fissility are interpreted as being formed by the collapse of a flocculated clay card house structure through processes of consolidation following deposition of large clay concentrations in seawater during sedimentation (O'Brien 1970b). More recently, O'Brien and Hisatomi (1978) differentiated depositional sequences in a turbidite cycle on the basis of fabric characteristics. The deposits attributed to rapid and concentrated turbidite currents were characterized by randomly oriented clay flakes. In contrast, deposits thought to have been deposited from more dilute suspensions and under slower sedimentation rates were characterized by parallel-flake orientation. Readers are referred to a paper by Heling (1970) in which the structure of shales and particle rearrangement during consolidation are discussed.

In many studies that address the possible factors responsible for fissility or nonfissility in shale, researchers have explored this interesting problem in terms of physico-chemical interactions and mechanical changes due to consolidation processes (White 1961; Gipson 1965, 1966; Odom 1967; O'Brien 1968, 1970b; Heling 1970). O'Brien (1981) has discussed the use of SEM in studying shale fabric; excellent photomicrographs depict the highly oriented fabric of fissile shales and the randomly oriented clay particles of nonfissile massive argillaceous rock. Byers (1974) pointed out the apparently long-neglected aspects of infaunal activity in soft sedimentary deposits. He showed that for some shales, fissility increased along a gradient of decreasing bioturbation. Massive rocks showing virtually no laminations were found to be characterized by a largely bioturbated fabric. Byers (1974) concluded that the presence of a very large carbon content in some shales resulted from lack of aerobic decomposition of organic detritus in muds. Apparently, black shales are commonly the most fissile (Ingram 1953). In future studies, more attention should be given to biological activity as a possible factor in the ultimate fissility developed in argillaceous rocks. In a study of pre-Quaternary clay sediments, Pusch (1971) showed that preferred particle orientation was apparent in clays subjected to the loading stress caused by Pleistocene glacier ice. He suggested that domains observed in some of the sediments may have been formed by the squeezing of clay particles during consolidation under large stresses.

In sharp contrast with the earlier concepts of single-clay-particle fabric, there was a pronounced increase in the observations and studies discussing the presence of compound, multiplate particles, such as domains, aggregates, and crumbs. These observations were made by soils scientists, as well as engineers and geologists (Pusch 1970; Barden et al. 1970; Burnham 1970; Green-Kelly and Mackney 1970; Yong 1972; Barden 1972; Smart 1972; Moon 1972; Rao 1972; Bennett and Bryant 1973; McConnachie 1974; Collins and McGown 1974). The importance of domains and compound multiplate particles, in general, was recognized as a significant aspect of clay fabric. The influence of compound particles on the engineering behavior of sediment was studied by several of the above investigators. In most of the current studies of clay fabric, the presence of multiplate particles in various sediment types is recognized.

Yong (1972) discussed the concept of the mechanics of crumb-particle interaction and particle orientation under load. The crumb is considered to be equivalent to the aggregate showing finite particle shape and structural integrity (fig. 3.23). Steps in particle-crumb interaction under load are as follows: (1) Rearrangement and reorientation of crumbs (aggregates) occurs without significant distortion within crumbs. (2) Further loading accentuates reorientation. (A type of pseudo-anistropy may result from alignment of the crumbs at this point.) (3) Crumb distortion and volume change occur at high stresses. Although Yong (1972) claimed that this sequence can be confirmed by fabric analyses, he did not assign specific stress conditions (specific loads) to the pro-

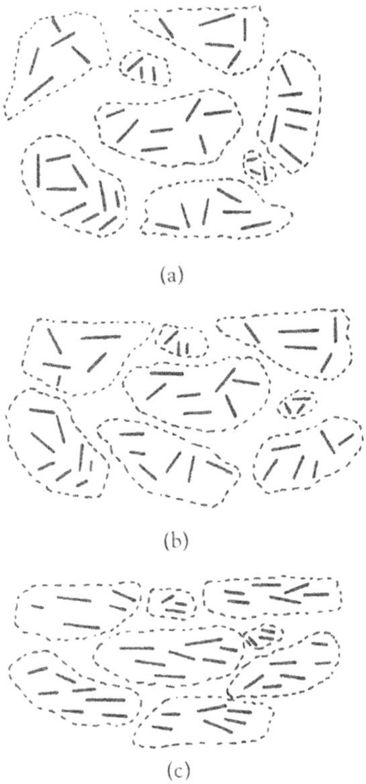

Figure 3.23 *Crumb structure and particle orientation under load (redrawn from Yong 1972): (a) Initial stress-initial reorientation of crumbs. (b) Further stress—accentuated preferred orientation of crumbs. (c) High stress—orientation of crumbs and particles within crumbs.*

cess during loading. This concept, however, is important in that at small loads, crumbs or aggregates are considered to respond initially with little or no single particle reorientation, and at larger loads, the crumbs distort, and single particles begin to reorient in response to the imposed stress condition.

The microstructure of artificially dispersed and flocculated kaolinite, illite, and montmorillonite was investigated (Sides and Barden 1971). Dispersed samples were prepared by using sodium oxalate, consolidating them with a uniaxial load of 29.4 kPa, and air drying. Calcium hydroxide was the flocculating agent, and these samples were consolidated under the same load as the dispersed samples. Dispersed kaolinite had a well-oriented structure, whereas flocculated kaolinite had orientated grains; there was a tendency towards horizontal bedding, and the samples possessed a turbostratic structure. Samples prepared from dispersed illite showed well-oriented grains, and some flakes formed a "skin" around silt particles. This skin was described earlier by Brewer (1964) and Burnham (1970). The flocculated illite structure was definitely random, but there were some local areas of oriented particles. There was a marked scarcity of single edge-to-face contacts indicative of a cardhouse fabric. The overall structure was simply described as *salt flocculated.* Sides and Barden (1971) found that illite was affected more strongly by dispersing and flocculating agents than was kaolinite. The effects of chemical additives on montmorillonite were similar to their effects on illite, but details of the structure were impossible to distinguish. Montmorillonite particles curled at their edges; this could have been caused by preparation techniques. Yong (1972) showed that if dehydration of the sample is not performed correctly, the fabric obtained is misleading and misrepresented. When montmorillonite or bentonite is air dried, the fabric produced is the so-called cornflake structure of curled clay platelets. This effect also was shown earlier by Sloane and Diamond (1970).

Sides and Barden (1971) concluded that as clay particle size is reduced, chemical additives become more effective in producing dispersed or flocculated structures. Thus illite showed greater effects due to chemical additives than did kaolinite, and no definite conclusions were reached for montmorillonite, because its extremely small size renders interpretation difficult. Bhaskaran (1972), in a reply to Sides and Barden (1971), stated that the degree of orientation of clay particles is a function of the type of clay particles, percentage of chemical agent present, its influence on the clay, and the consolidation pressure. Kaolinite tends to align into parallel arrays with less resistance than do other clays, and kaolinite may require a different quantity of flocculating agent in order to show significant effects.

Rao (1972), also replying to Sides and Barden (1971), pointed out that calcium hydroxide was a very poor choice of flocculating agent in studying clay fabric, because it attacks certain clay minerals (kaolinite and montmorillonite), causing hydrolysis of Si-O-Si bonds in the silica tetrahedron, with the formation of films of calcium silicate compounds at the particle surfaces. Obviously, strong chemicals used to treat and prepare clays for fabric studies may produce unnatural physico-chemical conditions and questionable fabrics compared with naturally formed clay fabrics.

An excellent study was made of the microstructural changes in a marine quick clay at failure (Pusch 1970). Electron micrographs revealed that the clay was characterized by a very porous network of dense aggregates connected by linking particles (fig. 3.24). During initial failure (in unconfined compression tests), link distortions developed. Aggregates appeared to move as units in connection with deformation of their links. At relatively large shear stresses, motion of the aggregates caused large deformations of connecting links. This resulted in parallel orientation of the linking particles (fig. 3.25). Remolding caused successive breakdown of aggregates and a decrease in shear strength. According to Pusch (1970), the shear process probably is governed by the action of the aggregates, and their rigidity may be due to the small distance between individual particles. The greater viscosity of water surrounding an aggregate relative to the viscosity of the free pore water, together with the strong bonding of the closely spaced particles composing an aggregate, result in considerable rigidity. A fabric characterized by a linkage of groups or chains of small particles between flocs and aggregates was found earlier by Pusch (1966b) in a quick clay.

In an excellent detailed study, Pusch (1973b) investigated the influence of organic matter on the geotechnical properties (strength, sensitivity, and consolidation) of clays. Using transmission electron micrographs, Pusch showed the strong and apparent affinity of organic material for clay particles. Clay aggregates and generally a more porous (i.e., large voids) arrangement of clay particles were

Figure 3.24 *Structure of a marine quick clay depicting aggregates and links prior to compression (shear deformation) (redrawn from Pusch 1970).*

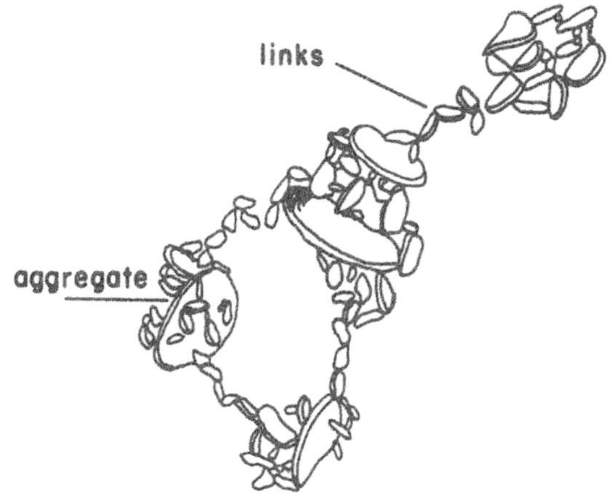

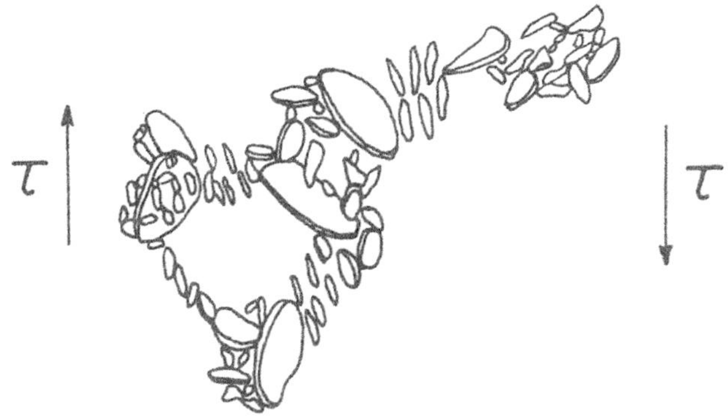

Figure 3.25 *Structure of a marine quick clay, depicting breakdown of links, resulting in domain formation and orientation of linking grains (same as figure 3.24 after shearing) (redrawn from Pusch 1970).*

apparent in marine clays. Dense aggregates and local humus bodies and particle links were present. In contrast, freshwater clays were characterized by local humus bodies in relatively porous aggregates but with overall small voids. Although little is known of the behavior of the complex integrated clay-organic system, Pusch (1973b) summarized three important aspects:

1. Organic species and residues in the form of large organic molecules and gel complexes can block small pores, thus producing dense particle aggregates of very small permeability.
2. Plant tissues and microorganisms, such as threadlike plant fibers, hyphae, flagella, ciliae, and some bacteria, are flexible and can be deformed without rupture. These substances produce a microstructural network of integrated clay-organic complexes characterized by large deformability, which may behave in an elastic, viscous, or plastic manner.
3. Organic gel complexes of humic substances may produce somewhat strong but largely deformable clay-organic aggregates. In the marine environment, where the electrolyte content is large, clay particles and organic substances collect into dense aggregates, whereas in freshwater clays, the aggregates are less dense and more deformable.

Studies of the New Liskeard varved clay from Ontario revealed a flocculated open soil structure. This structure would not be expected for deposits from cold, fresh melt waters of glacial lakes; a dispersed condition would be more likely (Quigley and Ogunbadejo 1972). The clay possessed a preconsolidation pressure that was greater than overburden pressure. Relatively slow sedimentation rates, bond development (such as cementation by carbonates), and significant van der Waals forces were considered to be responsible for the overconsolidated characteristic. Comparison of the New Liskeard clay to soft stratified glacial sediments of similar compositon from Welland, Ontario, indicated that the two clays differed in physical properties. Platelet parallelism was attributed to more rapid sediment accumulation and compression rates, together with an insignificant degree of bond development for the Welland clay.

Kirkpatrick and Rennie (1972) reported on the fabric of laboratory-consolidated kaolin as observed by electron microscopy. The samples were air dried, which may be expected to alter significantly the sample fabric. However, samples that had been subjected to different consolidation conditions before drying did show different fabrics. These differences provide at least weak evidence of genuine differences in fabric resulting from different consolidation conditions. The results indicated that kaolins consolidated under isotropic condition possessed an isotropic structure, with the clay particles oriented at random. Kaolin samples consolidated anisotropically possessed a strongly oriented structure, with the clay particles oriented at right angles to the direction of the principal stress.

The increasing observational evidence suggests that single-clay-particle fabric of sediment is not wholly tenable, and sediment fabric is more accurately represented by domain, aggregate, and linking-chain particle arrangement (Bennett et al. 1977). Models of clay fabric that are beginning to emerge incorporate these more recent observations and concepts. Moon (1972) proposed a model of clay particle arrangement for unconsolidated versus consolidated sediment, representing flocculated and dispersed conditions. Conceptually, the model accounts for the basic domain particle interaction during consolidation, with particle alignment of domains in preferred orientation (fig. 3.26). As consolidation proceeds, more particles are incorporated into each domain. Domains are of the book and stepped face-to-face types.

Fabric unit interaction and classification have been discussed in order to predict responses of soil and to develop fabric models that more closely approach a physical description of real sedimentary material (Yong and Sheeran 1973). Fabric units were defined as groups of particles. Thus the fabric unit includes crumbs, aggregates, clusters, floccules, and particles, which are grouped into domains. Yong and Sheeran stressed the importance of recognizing that in most cases, the fabric unit consists of multiple particles, rather than single particle arrangements making up the basic structure of sediment.

Current concepts of clay fabric were presented at the International Symposium on Soil Structure held in Gothenburg during 1973. The meeting had four basic themes: (1) Techniques of observation and methods of quantification, (2) macrostructure and microstructure of soils, (3) physico-chemical processes that affect soil structure and the impact of soil structure on local physico-chemical processes, (4) mechanisms of deformation and failure.

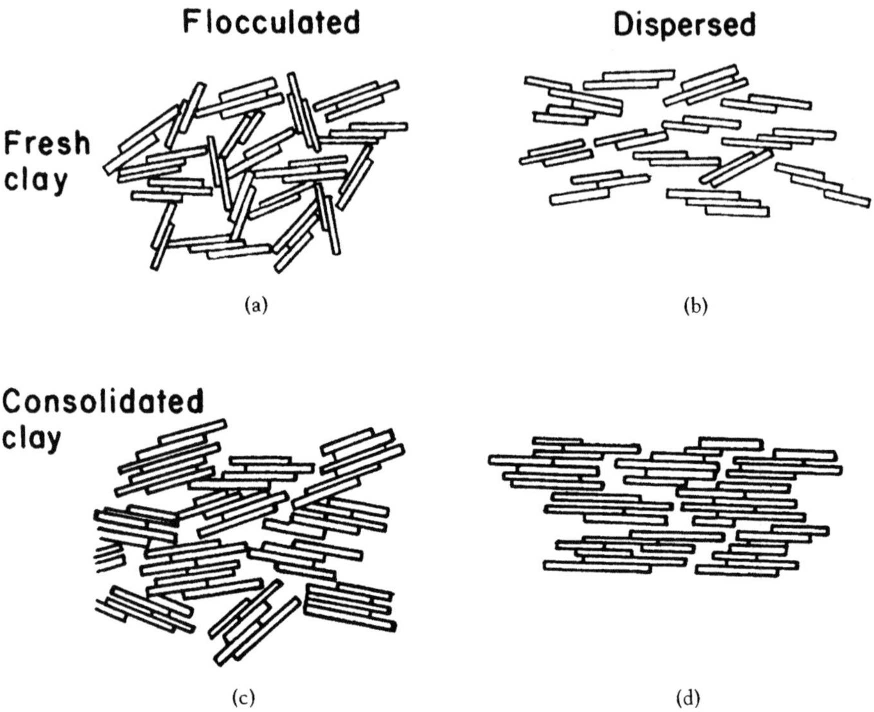

Figure 3.26 *Proposed schemes of particle arrangement in clay sediments (redrawn from Moon 1972): (a) Open, random arrangement of domains of 2 to 3 particles per packet. (b) Parallel or subparallel arrangement of domains of 2 to 3 particles per packet. (c) Increased parallelism of domains and more particles incorporated into each domain than in (a) (mudstone). (d) Complete parallelism and more particles per packet than in (b) (shale).*

Barden (1973) summarized the various factors important in the development of soil structure:

1. The electrochemical environment at the time of deposition, such as that of hypersaline, marine, brackish, or fresh water. This factor has the dominant influence on the structure.
2. Changes in stress during loading and unloading that lead to the collapse of initial structures and the development of preferred orientation of the fabric.
3. In some cases, the effects of chemical and physical weathering may be strongly disruptive and remove the effects of overconsolidation.

Barden (1973) also summarized important general conclusions that have resulted from recent observations of sediment microstructure. Single-particle behavior is very rare, and the cardhouse particle arrangements are theoretical idealizations of fabric rather than a common characteristic. Clay plates usually are aggregated in some fashion; this observation confirms Yong's suggestion that the single plate theory may be relevant to dilute colloid suspensions, but it does not apply generally to natural clays. This concept of multiple plates, according to Barden, may suggest some resemblance between clays and granular soils.

Flocculated structure may result following sedimentation in marine, brackish, or fresh water, and the degree of openness of the fabric is influenced by the mineralogy, clay plate size, and amount and angularity of the silt particles present. Consolidation causes collapse of the bookhouse fabric into a turbostratic fabric, possibly causing horizontal orientation. Arching effects can maintain areas or pockets of open-flocculated structure.

The single factor that most influences the final structure of a clay is considered to be the electrochemical environment during sedimentation and its effect in establishing a flocculated or dispersed structure. Although there are many factors involved in the development of soil structure, Barden claims that the study of microstructure is very important in understanding and linking geological processes involved in the development or formation of the soil (sediment) and its geotechnical properties.

Smart (1973) has reviewed comprehensively the statistics of fabric analysis as observed by electron microscopy. He discussed various methods of statistical analysis of particle size and orientation in both three-dimensional and two-dimensional cases for clay fabric. Tovey and Wong (1973) and Lisle (1976) extended the discussion of quantitative descriptions of fabric. They emphasized the importance of recognizing, in quantitative studies, the possibility of the introduction of bias into the results by observation of too few specimens or by observation of specimens too limited in area.

A study was published on microfabric features, their form and function, in various natural sediments (Collins and McGown 1974); including marine water, brackish water, fresh water, floodplain, lacustrine, aeolian, and glacial deposits. The objectives were to delineate the particular types of fabric, to establish whether or not earlier concepts of particle arrangement were realistic, and to find out whether or not the specific fabrics were associated with a specific engineering behavior of the sediment. SEM was used to investigate the fabric, and the samples were prepared using the air-dry-fracture-peel-coat techniques described by Barden and Sides (1971). Although the air-dry technique is suspect in preserving the fabric of undisturbed sediment, the investigators apparently assumed that this factor was not critical. This study is one of a few investigations that have addressed questions of environment and fabric. Thus a review of the general findings will be given here, despite the fact that the reported results may not be entirely representative of the undisturbed natural sediment. The relations between fabric and depositional environment deserve more attention in future investigations.

Collins and McGown (1974) claimed that the widespread assumptions of unique relations between microfabric, mode of deposition, and depositional environment were not substantiated by their study. They found, however, that there may be a dominant fabric feature or set of features characteristic of any one sediment. Commonly, different types of microfabrics were observed in a single type of sediment. The dominant microfabric features characteristic of particular sediments were associated with engineering properties, such as sensitivity, collapse, and expansion. The medium-sensitive to extra-quick sediment was characterized by particle arrangements of both bookhouse and stepped flocculated fabrics, and a type described as *partly discernible particle arrangements*. Turbostratic arrangements were observed in the medium-sensitive sediment. All sensitive material contained clay matrices. The very sensitive quick clays from Norway often possessed connector and irregular aggregation assemblages. Linkages between aggregations were composed of a number of the connector assemblages. Other medium-sensitive clays from various localities showed interweaving bunches that tended to be oriented toward the horizontal. Aggregation assemblages were observed in some sediments of medium sensitivity. Connector assemblages were very infrequent in the less sensitive sediment.

Collapsing sediment, usually composed of silt, clay, sand, or till, contained elementary particle arrangements of fine silt and clay particles in various configurations. The partly discernible type fabric also was present in these sediments, and the onionskin fabric, which the investigators called *clothed grain-grain* was observed. Connector assemblages were common in these sediments, and the granular particle matrices were predominant. Regular aggregations were observed in the collapsing sediments, but they were rare in the loess material. Collins and McGown observed that in the expansive sediment, the predominant microfabrics were the clay particle matrix assemblages, although fabric types included the connector fabric and honeycomb fabric and the clothed grain-grain contact features.

Various types of fabric were observed by Collins and McGown (1974) in sediments. Specific types of fabric occurred in numerous sediments, but certain fabric seemed to be characteristic (i.e., predominant) of certain types of sedimentary material. The study pointed out that many of the particle arrangements and assemblages present in natural sediment have not been observed in laboratory-prepared monomineralic clays, and laboratory test specimens may not possess the same fabric as that of natural sediment. The study also included a description of the pore space of the sedimentary material, which followed closely the classifications of Brewer (1964) and Bochko (1973).

As an integral part of the study of microstructure in general, several investigators have focused on the characteristics of pores and pore systems in sediments and rocks (Diamond 1970; Timur et al. 1971; Sarkisyan 1971; Bochko 1973). The scanning electron microscope has been a basic tool for much of this research.

The size, shape, geometry, and other characteristics of pores, such as interconnection, are being studied because they are important in petroleum recovery and in determining rates of fluid flow through sediments. Pittman and Duschatko (1970) discussed the use of pore casts and scanning electron microscopy to study the pore geometry of rocks. A very intricate pore network was observed in quartz sandstones with low porosities. Techniques of image analysis, computer graphics, and optics have been developed in order to quantify pore geometry for the study of sandstone (Lin and Hamasaki 1983). These techniques, although developed for the study of coarse-grained clastics, may hold promise for the study of fine-grained sediments. The techniques described by Lin and Hamasaki have been applied to geological materials of low porosity and permeability: both properties are characteristic of consolidated muds. During an investigation of the permeability of deep-sea clays recovered from the East Bermuda Rise and Blake Bahama Outer Ridge, Clukey and Silva (1982) showed that the sediment had a flocculated fabric characterized by evenly distributed clusters and voids. Clukey and Silva contended that the fabric was similar to the microfabric of saturated clays described by Olsen (1962). All three authors discussed the importance of flocculated microfabric in controlling the permeability of fine-grained sediments. Attempts have been made to apply corrections to the Kozeny-Carman equation for predicting the coefficient of permeability, but considerable quantitative research remains to be done for a clear understanding of the relations between fabric and permeability.

Smart (1975) has written an interesting, brief review on soil microstructure, with particular reference to clay fabric and the mechanical behavior of soil. Smart reviewed definitions of several terms and discussed aspects of clay fabric of soils, such as pores, fabric arrangements (open structures and dense structures), stress effects, consolidation, deformation and failures, and composition. He also pointed out the probable importance in some cases of pore size in relation to functional characteristics, such as sizes available to bacteria and free-draining pore sizes. In these respects, Smart discussed a classification of pore sizes ranging from 6 mm to 0.2 μm; it has three major divisions—mini, macro, and micro, from largest to smallest, respectively. He discussed briefly the possible stabilizing effect of bacteria on soil structures and bacterial rearrangement of particles near surfaces of pores. It is interesting that later Maclean and Smart (1978) showed by TEM ultrathin-section micrographs a possible bacterium surrounded by a well-defined microstructure.

McConnachie (1974) reported on fabric changes induced by laboratory consolidation of monomineralic kaolin, which was prepared by mixing at a moisture content of 250%. The original void ratio (ratio of the volume of the voids to the volume of the solids) was about 5.5. The material was consolidated over a range of about 0.1 to 100,000 kPa. A strong inflection in the void ratio versus the logarithm of pressure occurred at 14.7 kPa, creating two linear portions of the curve. The original clay fabric was characterized by edge-edge and edge-face contacts among domains. As the pressure increased, the domains rotated and became more closely packed, and at approximately 14.7 kPa, the domains became reduced in size. This change in domain size was attributed to a slippage of particles in face-to-face contact during rotation of the domains as consolidation proceeded. The change in behavior of the kaolinite domains at the pressure of 14.7 kPa is attributed to the collapse of the clay fabric, and it is considered to correspond to the greatest pressure at which interdomain and interparticle physico-chemical forces can be sustained. The fundamental change in the mechanism of consolidation occurred between 9.8 kPa and 98 kPa. Apparently, most of the increase in particle orientation was complete at about 9.8 kPa. The domains moved closer together at greater pressures, but without significant change in orientation. At about 100,000 kPa, the domains showed a degree of orientation that was far from perfect, with the mean angle of 22° between domains. Apparently, the greatest degree of particle orientation occurred in the lower pressure level, and little increase in orientation occurred even at very high loads. The sediment investigated, however, was relatively pure kaolinite that was prepared at a specific water content in the laboratory. Thus the fabric and material engineering behavior may not be representative of naturally occurring sedimentary material. Studies of high-pressure consolidation of kaolinitic clay revealed that preferred orientation of the clay particles develops at the drained end of the samples, whereas preferred particle orientation was prevented by the presence of high pore pressures at the undrained end of the samples (Wijeyesekera and DeFreita 1976). This study pointed out the importance of pore fluid pressure control during laboratory consolidation. Changes in pore fluid pressure as a function of time are a governing factor during consolidation of naturally occurring sediment.

The importance and effect of pH and electrolyte on the mode of clay particle (kaolinite) interaction in aqueous suspension was investigated by Rand and Melton (1977). Rheological properties of aqueous kaolinite suspensions were interpreted in terms of the fabric of the suspended floccules and nature of the particle contacts. Changes in the Bingham Yield stress as the hydrogen ion concentration was altered were suggested to be controlled by changes in the fabric of the clay floccules (fig. 3.27). Clearly, the role of physico-chemistry in relationship to clay fabric, particularly in clay suspensions, cannot be overemphasized. Physico-chemical aspects of clay fabric have received the attention of numerous researchers in the field of colloidal chemistry.

Pusch (1976) addressed aspects of shear deformation of the clay microstructures of soils in detail. Pusch showed that it is inappropriate to make larger-order assumptions in applying the rate process theory to time-dependent shear deformation (specifically to effects at particle bonds). Microstructural investigations have revealed that clay particle networks consist of relatively rigid aggregates connected by weak links of particles commonly called chains. This type of microstructure results in greatly variable stress levels under given loads, in direct conflict with rate process theory assumptions of soil particles uniformly distributed and equally stressed under specific loads. Pusch pointed out that only a very small fraction of the total number of particle bonds become activated fully even at stresses that produce bulk failure. He stated that the degree of strength mobilization is highly variable in a clay element subjected to loading, cohesion is equivalent to the integrated strength of the links, and internal friction is due to viscous effects and dilation. Pusch (1976, 270) stated that these factors are "the very essence of the modern concept of clay shear strength."

Bennett (1976) and Bennett et al. (1977) investigated the clay fabric of submarine sediment from selected cores recovered in the Mississippi prodelta and Pacific Basin. The delta and Pacific Basin samples cored to depths of 152 and 144 m below the mudline, respectively, were characterized by a predominance of smectite and illite clay minerals. The transmission electron photomicrographs of the Mississippi Delta submarine sediments revealed clearly a clay fabric of randomly oriented domains characteristic of the mod-

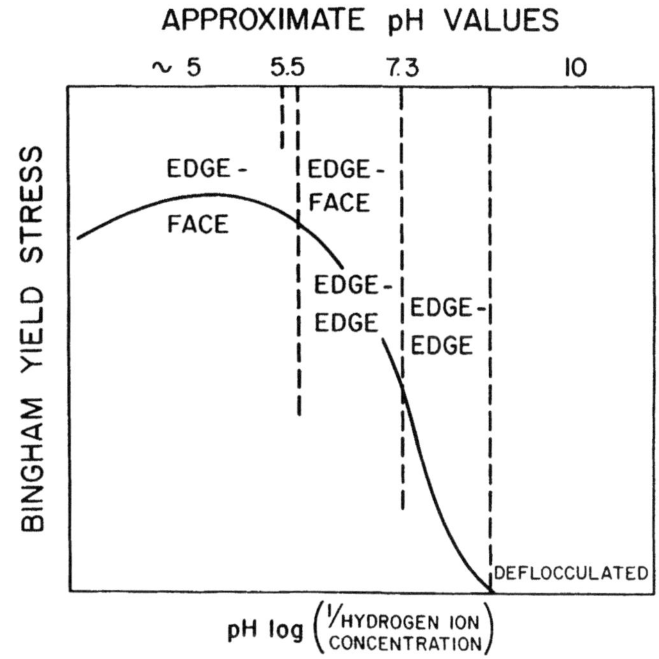

Figure 3.27 *The effect of pH on the Bingham yield stress of kaolinite suspensions in relation to the mode of particle interactions (redrawn from Rand and Melton 1977).*

erately high-porosity samples (porosity = 60–70%) to depths of approximately 37 m below the mudline. Domains were the predominant clay particle type observed with some recognizable thin single-platelike particles associated particularly in close proximity to the large particles. The voids of the high porosity (upper) clays ranged in size from small to relatively large and their shapes were irregular. The deeply buried, relatively lower-porosity (<55%) clays were characterized by a strongly oriented clay fabric with associated voids which were thin and long. A dominant preferred particle orientation, but not complete preferential arrangement, was found to be the rule for these lower-porosity clay sediments.

Void ratio (e) and porosity (n) are fundamental properties of sediment which readily reflect changes in water content with increasing depth of burial through the processes of dewatering and consolidation. Porosity is related to void ratio by

$$n = e/(1 + e). \tag{3.1}$$

Changes in void ratio of a given sedimentary deposit of relatively uniform grain size distribution and uniform grain density reflect clearly the corresponding changes in the basic mass physical proprties, such as wet bulk density and water content. Void ratio is an excellent property for comparing samples of a given sediment (particularly with reference to consolidation, dewatering, depth of burial, and fabric), because the voids decrease in size with increasing overburden stress, in contrast to the solids, which maintain a constant volume acting as a common denominator for comparison purposes. Thus void ratio indicates a direct relationship between the solids and the voids of a given sediment mass.

Decreasing water content, void ratio, and porosity and increasing bulk density as a function of depth below mudline is an accepted axiom of marine geology and sedimentology. Although exceptions and variation to the rule do occur, the basic observation of decreasing water content with increasing sediment depth, particularly for a specific sediment type, is essentially ubiquitous for submarine sediments. Numerous electron photomicrographs of the clay fabric from Mississippi Delta cores were examined with reference to depth of burial and have revealed clearly a decrease in void ratio, change in general voids shape, and particle rearrangement (Bennett et al. 1977, 1979).

Figure 3.28 *Void ratio versus depth in Mississippi Delta sediment. Values were either predicted by the Hamilton (1959) technique or measured in the laboratory.*

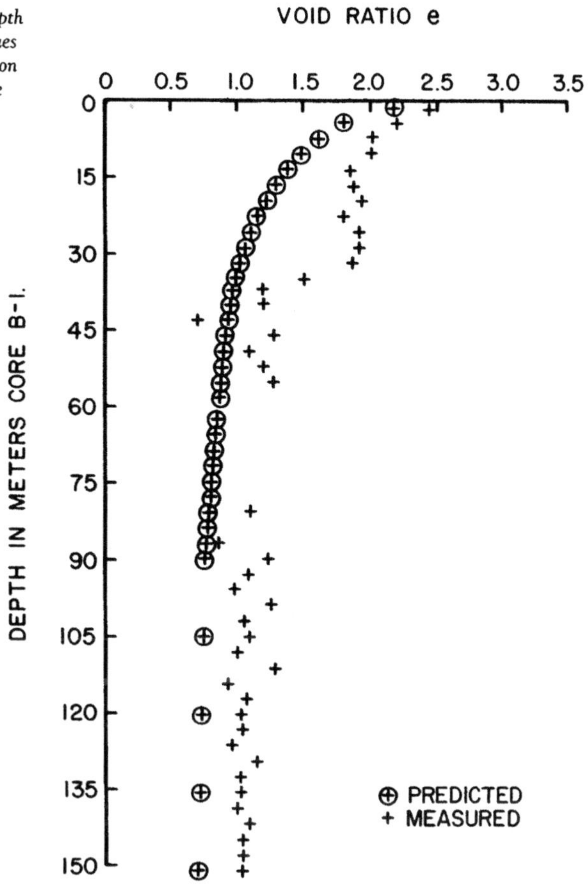

The void ratios were calculated for core B–1 (fig. 3.28) based on the measured mass physical properties such as water content, wet bulk density, and average grain density. These mass physical properties were determined by standard laboratory techniques. The predicted void ratios versus depth of burial were determined by the method described by Hamilton (1959), which he had used for the purpose of predicting the consolidation behavior of submarine clay sediment. The method is based on data determined from typical consolidation, e–log p curves from laboratory tests. The details of the method can be found in Hamilton (1959). Plots of void ratio versus depth of burial for both measured and predicted cases depict clearly a very close similarity in curve shape and change in void ratio with depth (fig. 3.28). Details of the analyses can be found in Bennett et al. (1977).

The change in void ratio with increasing depth of burial for the sediment investigated can be related to the characteristic clay fabric. Sediment of relatively high void ratio (fig. 3.28) was found by Bennett et al. (1977, 1979) to be characterized by a fabric of randomly oriented domains and short chains with particles in edge-to-face and some face-to-face contact (figs. 3.29, 3.30, 3.31). Sediment of intermediate void ratio shows greater particle-to-particle packing but a consistent predominance of randomly arranged particles with a few chains and perhaps a slight development of larger domains than observed in the high void ratio sediment (figs. 3.32 and 3.33). The low void ratio of the sediment, determined by the highly oriented clay particles and the thin, long, narrow voids, is brought about by greater density packing of the particles into long chains (fig. 3.34). The void geometry is markedly different between the high void ratio surficial sediment and the deeply buried, low void ratio material. This is a result of major particle realignment due to dewatering and consolidation processes.

Through the application of image enhancement and digital image processing of the delta high-porosity sediment, Bennett (1976) and Bennett et al. (1977) confirmed the presence of multiplate particles or domains. The particular clay sample analyzed was buried under only 1.4 m of relatively recently deposited sediment, with obviously minimal overburden stress ($\cong$ 12 kPa) imposed on the material.

The Pacific deep-sea clay fabric sample (from Deep Sea Drilling Projects (DSDP) leg 16) was characterized by flocs and linking

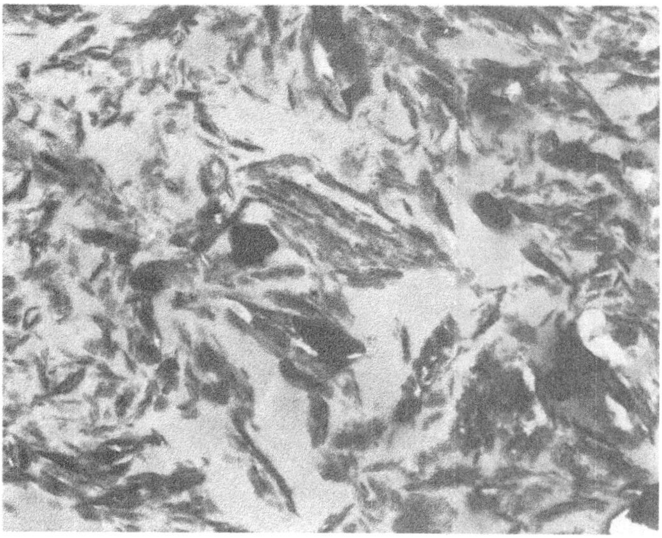

Figure 3.29 *Clay fabric of randomly oriented domains (Mississippi Delta, 1.4 m below seafloor, TEM). (Magnification 1600X)*

Figure 3.30 *Typical fabric of high void ratio Mississippi Delta surficial sediment (SEM, 0.3 m below seafloor), showing numerous edge-to-face contacts. (Magnification 1200X)*

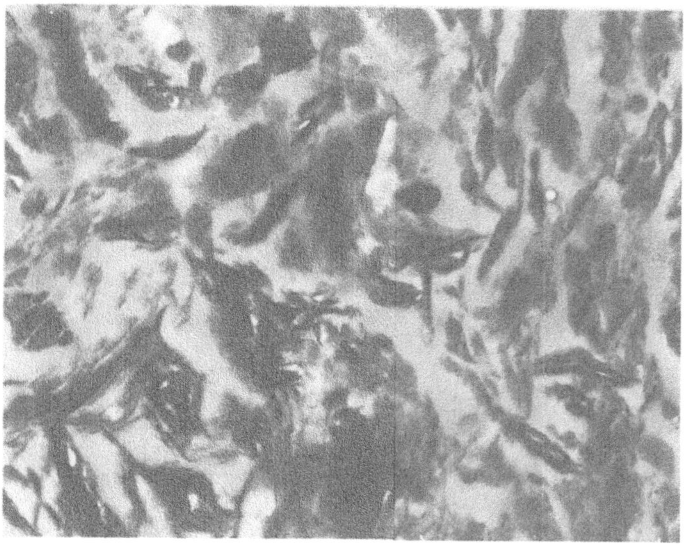

Figure 3.31 *High void ratio sediment characterized by randomly arranged domains (2 m below seafloor) and showing edge-to-face particle contacts (TEM). Compare with figure 3.29. (Magnification 1600X)*

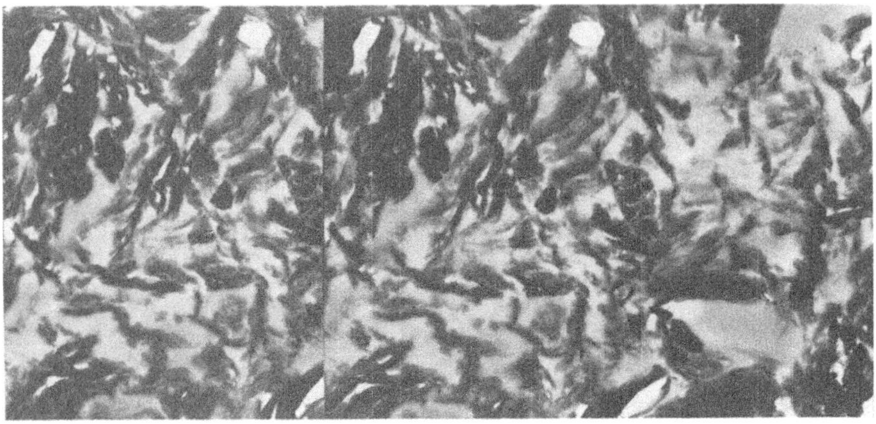

Figure 3.32 *Clay fabric of relatively large voids and randomly oriented domains (27.7 m below seafloor, TEM-stereo). (Magnification 12,750X)*

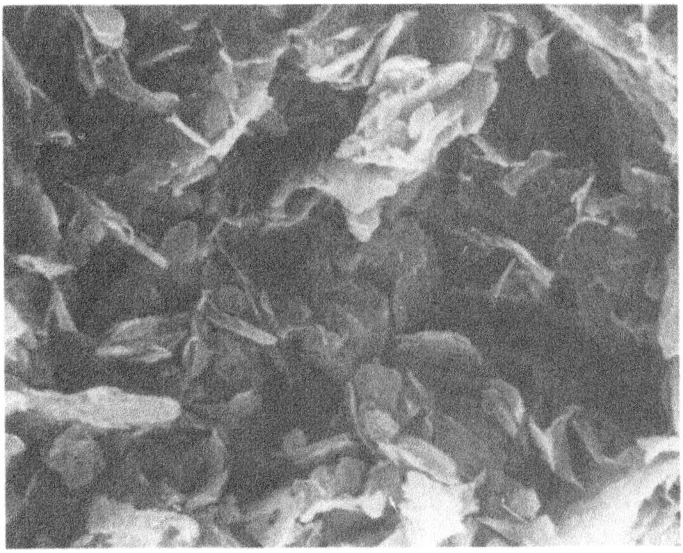

Figure 3.33 *Clay fabric of randomly oriented domains, showing particle contacts and large voids (26.5 m below seafloor, SEM). (Magnification 10,000X)*

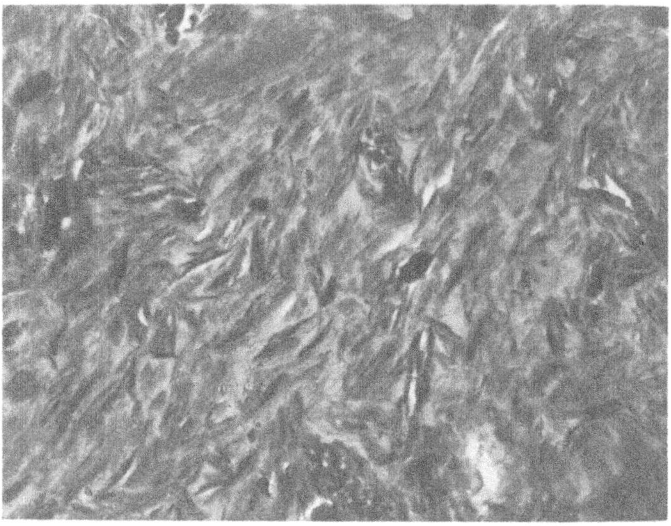

Figure 3.34 *Clay fabric of domains strongly oriented by natural consolidation processes. Not all of the particles are oriented preferentially (144.8 m below seafloor, TEM). (Magnification 16,000X)*

chains (figs. 3.35 and 3.36), which contrasted sharply with the fabric of the Mississippi Delta samples. In addition, the void ratio of 3.5 and porosity of 78% determined for the DSDP sample differed markedly from the delta samples. The clay fabric of the DSDP sample, however, resembles closely the observations and model depicted by Pusch (1966, 1970) by virtue of the flocs and linking chains observed in his TEM studies (fig. 3.24). This conclusion was corroborated by Pusch (personal communication 1979).

Following the clay fabric observations, Bennett (1976) and Bennett et al. (1977) prepared tentative models of clay fabric for submarine sediments composed predominantly of smectite and illite. The models were related to characteristic void ratios of the sediment and all models proposed, with the exception of a low void ratio model ($1.2 < e < 1.5$), have been observed for various clay-rich submarine sediments (fig. 3.37).

The major conclusions regarding the clay fabric observed for the Mississippi Delta samples are (1) domain particles do exist in these moderately high porosity, only slightly consolidated, submarine clay sediments, (2) strong preferred particle orientation is achieved at depths of not greater than 120 m and probably somewhat less. The DSDP high porosity, high void ratio, deep-sea clay fabric was characterized by flocs (with large intravoids) connected by linking chains, thus producing a sediment of high water content even at depths of 140–144 m below the mudline. As pointed out by Bennett (1976), the difference in the fabrics of the two sediment types is caused clearly by the significant differences in depositional environment, sediment source, and related environmental conditions. Rates of sediment deposition and source of supply are quite likely important contributing factors responsible for the differences in the ultimate clay fabrics of the two sediment types.

A deep-sea core from the Vema Channel was investigated in detail by Faas and Crocket (1983) over its 100-m length. Fabric analysis and studies of mass physical properties showed that the surficial sediments were characterized by randomly oriented domains, and increasing overburden with depth of burial resulted in a fabric of more parallel, face-to-face arrangement of particles. The analysis by Faas and Crocket demonstrated that preferred particle orientation in the deep-sea clay deposit developed during normal consolidation at overburden pressures of approximately those found by Bowles et al. (1969). The results substantiated the fabric development model described earlier by Bennett et al. (1977, 1981).

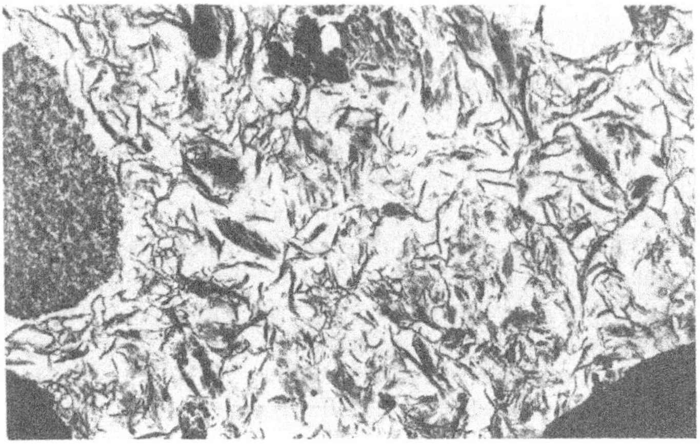

Figure 3.35 *Clay fabric showing flocs, linking chains, and zeolitic aggregate (Deep-Sea Drilling Project, core 163A, 143 m below seafloor; TEM). (Magnification 10,000X)*

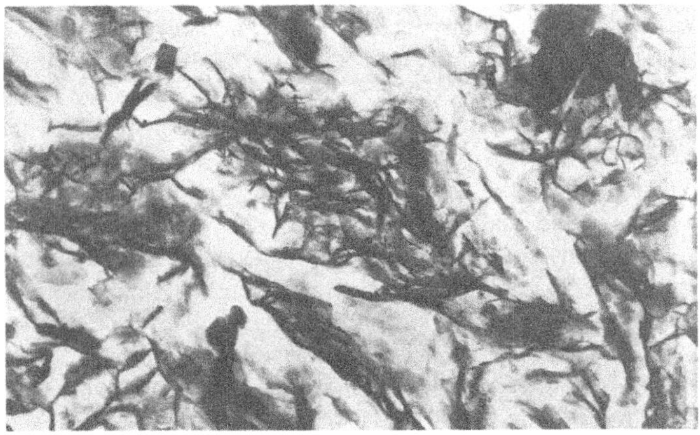

Figure 3.36 *High TEM magnification showing individual clay flocs, floc intravoids, and linking chains (Deep-Sea Drilling Project, core 163A, 143 m below seafloor). (Magnification 29,500X)*

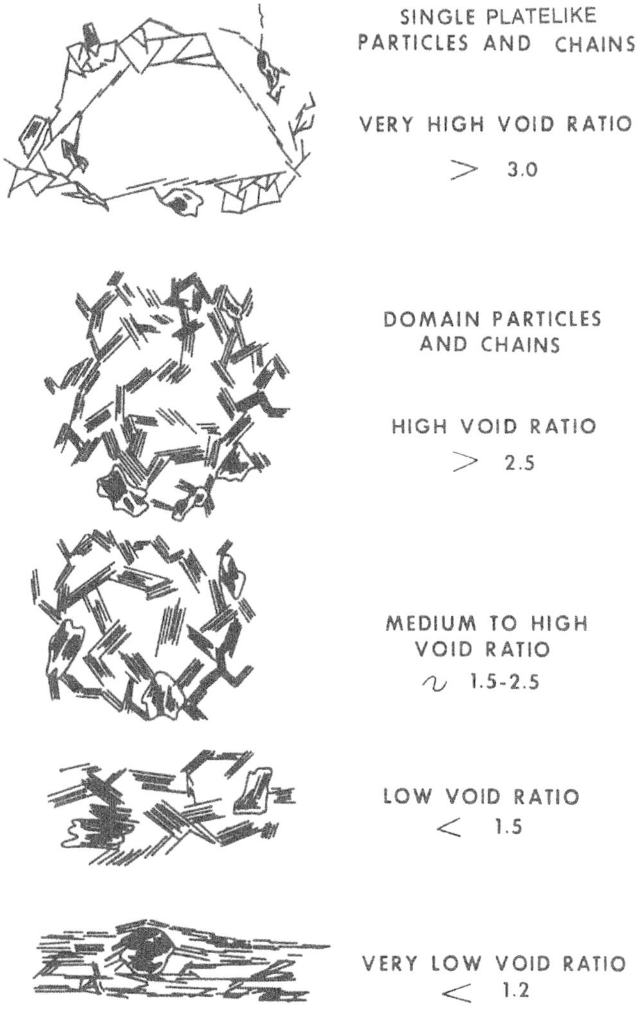

Figure 3.37 *Fabric models for submarine sediment. (From Bennett 1976; Bennett et al. 1981.)*

Advances in clay microstructure were presented at the Clay Minerals Society Annual Meeting and Conference in Waco, Texas, during 1980. Numerous presentations covered subjects dealing with fabric, geotechnical properties, and depositional environments (O'Brien 1980; Chiou 1980; Bennett et al. 1980). Other presentations dealt with clay microstructure as related to engineering applications (Veniale 1980; Barelli 1980). Fabric in relation to energy source reservoirs was discussed by Hower et al. (1980). They showed that the presence of clays in sandstone formation pores is significant in oil and gas production. Again, this points out the importance of clay fabric in influencing the fundamental physical properties of sediments. Further applications of the SEM to geological investigations were presented at the SEM Symposium in Arizona during 1981, reemphasizing the steadily increasing use of the electron microscope in the study of clay microstructure.

The microstructure of a Lake Superior clay was reported by Johnson (1983) to reflect the effects of ice movement. He concluded that glaciation produced linear streaks predominantly perpendicular to the ice flow and dipping in the up-ice direction by generating shear planes. The unexpected deposition of clay in a well-flushed glacial lake was attributed by Smith and Syvitski (1982) to aggregation of the fine-grained material in fecal pellets. The breakage of flocs by hydrodynamic shear was studied by Gibbs (1981) and the opposite process, the creation of flocs from dispersed particles by the action of hydrodynamic shear, was investigated by Koh (1984).

Calculations of potential energy surfaces computed for clay particle assemblages were shown to provide insight to changes in fabric that can occur when clay masses are subjected to chemical change or mechanical disturbance. It was shown that studies of potential energy surfaces can give important clues to understanding fabric development. The relative importance of the chemical environment and mechanical factors during initial and postdepositional stages of fabric development, respectively, were discussed (Hulbert and Bennett 1980).

Clearly, the importance of gaining a deeper understanding of clay microstructure has been recognized by numerous scientific and engineering disciplines. Clay fabric and physico-chemistry of fine-grained, mineral sediments cannot be ignored if (1) fundamental properties of sediments are to be understood more fully, (2) sedimentary materials are to be utilized more effectively, (3) the geological processes related to sedimentary deposits are to be elucidated clearly and related to the geological record, (4) the mechanical behavior of fine-grained sediments under static and dynamic loads are to be understood and predictable, (5) the acoustic wave propogation, acoustic impedance, and acoustic isotropy–anisotropy of geological materials from a variety of environments are to be explained and predicted reliably, and (6) the fluid flow (permeability) in sediments is to be understood and applied with confidence to practical problems.

4 A Current View of Clay Microstructure and New Research Directions

The accumulated literature has established that although numerous studies and advances have been made, a paucity of data and information exists in various important areas of research on clay microstructure. Only a few studies dealing with the fabric of natural sedimentary material have been reported, in contrast to numerous studies of laboratory prepared material. The only comprehensive studies of submarine sediment clay fabrics have been conducted by Bowles (1968a, 1968b) and others (Bowles et al. 1969), Bennett and Bryant (1973, 1976), and more recently, by Bennett (1976), Bennett et al. (1977, 1979, 1981), Bohlke and Bennett (1980), Guorui (1983), Faas and Crocket (1983). Very scant information exists on the initial stages of fabric formation in natural systems and the effects of organic and inorganic compounds on the development of clay fabric. Unfortunately, clay fabric samples for many TEM and SEM studies were not prepared properly, rendering the final observations suspect in terms of the original features characteristic of the sediment. Standards were found to be virtually nonexistent for clay fabric analysis, sample preparation, assessing bulk sample quality, and electron microscopy techniques. In a few studies, the need for careful laboratory techniques applied to the study of clay fabric was recognized. However, no basic standards have been established.

Preparation of Samples for Electron Microscopy

The most critical techniques in preparing samples for electron microscopy (EM) studies are in the dehydration of specimens that are fully or partially water saturated. Likewise, the techniques for embedding a sample with an appropriate medium, a necessary prerequisite to ultrathin sectioning, must be accomplished with great care. Appropriate care must be employed in order to insure excellent quality and reproducible results. Careful techniques will minimize sample disturbance and maintain basic structural integrity as well as particle-to-particle orientations in clay samples. Gillott (1975) presented a detailed treatment of the importance of specimen preparation techniques for microscopy investigations. Particularly in studies of clay fabric, the importance of preparation techniques cannot be overemphasized. Chiou (1980) described a new technique for clay fabric studies for the preparation of marine sediments at in situ pressure. This technique may hold particular promise for analyzing gassy sediments. Recently, Smart and Tovey (1982) completed a comprehensive compilation of techniques of electron microscopy for sediments and soils.

Dehydration and Embedding Methods

Studies have demonstrated that air- and oven-drying clay specimens severely distort the fabrics, rendering the structures of the samples unnatural (Yong 1972; Naymik 1974). Yong (1972) has shown that dehydration of monmorillonite clays by air-drying creates the "cornflake structure" that is observed commonly in photomicrographs. The clay flakes curl and warp (cornflake effect), due to surface tension at the fluid-vapor interface, which passes through the sample during drying. Naymik (1974) has shown clearly that the critical point drying technique is far superior to either air-drying or freeze-drying of clay-rich specimens. He found that a large reduction in pore space volume occurred during air-drying, and artifacts of drying formed during freeze-drying. He also reported an absence of desiccation shrinkage in samples that had been dried by the critical point technique. Obviously, indurated sediment would not be subject to the same seriously deleterious effect of drying as would weakly cohesive, high-porosity sediment.

Gillott (1969) discussed the details of freeze-drying and critical point drying of clay samples. He observed the strong likelihood that fabric would be affected by shrinkage during air-drying, as revealed by a large volume decrease when drying under ambient conditions. He attributed this change to the strong force of surface tension during air-drying, which causes clay plates to roll up and become tubular, thus seriously modifying the arrangement of the particles of weak clay sediments. Gillott (1969) also thought that artifacts may be formed by ice crystallization during freeze-drying, which damages the original fabric. Detailed studies of the freezing of physical systems have revealed that some amorphous portions, or bound water, do not freeze but are merely dried at low temperature. This constitutes a type of pseudo-freeze-drying (Luyet 1961). Luyet alluded to the fact that ice crystals form to some degree during the freeze-drying process. An example of gross changes in a sample is shown in figure 4.1, a photomicrograph of a clay sample which was frozen so slowly that significant crystallization of water occurred.

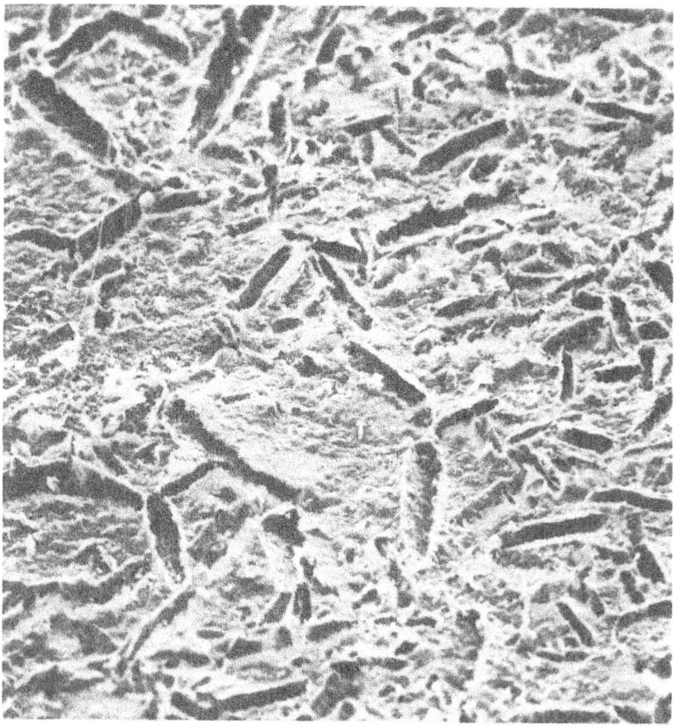

Figure 4.1 *Clay fabric of sample showing gross changes due to crystallization of water during freeze- drying (SEM, courtesy Neal Currier, International Minerals & Chemical Corporation). (Magnification 10,550X)*

O'Brien (1970d) described an interesting freeze-drying method for studying moist, unconsolidated, flocculated clay sediment. A drop of clay suspension is deposited on an aluminum disc and submerged in liquid isopentane for freezing. The sample is then freeze-dried, gold shadowed, and examined in the SEM. This technique of drying suspended, flocculated clays is undoubtedly better than air-drying procedures, owing to the extreme fragility of the flocs.

Other embedding techniques for studying soils by electron microscopy have been described by Pusch (1966b, 1967), Smart (1967a, 1967b), O'Brien (1971), Foster and De (1971), and Smart and Tovey (1982).

A detailed investigation of impregnating techniques was carried out by Foster and De (1971). Their results showed that the observed strains in the original fabric were minimal in soft material. Their techniques, however, did not involve drying; they involved a series replacement of the interstitial water by diffusion of various fluids to a final hardened resin. The technique is very time-consuming and laborious. Photomicrographs of submarine sediment prepared by freeze-drying, followed by embedding with epoxy resin and ultrathin sectioning, revealed clearly detailed, complex fabrics that appear to be unaffected significantly by sample preparatory techniques (Bowles 1968a, 1968b); however, adverse effects of drying were not delineated. The freeze-drying technique is also time-consuming, and there is danger of ice crystallization during the process. Delage et al. (1982) reported successful preparation of a sensitive clay by using a SEM specially equipped with an enlarged introduction chamber. A sample is frozen with liquid fluorocarbon at −145°C, fractured in the evacuated introduction chamber, metallized after surficial ice has sublimated, and examined in the cooled microscope column, which is still at a temperature below −100°C. Undoubtedly, prior to the development of the critical point drying technique, the freeze-drying method was the most satisfactory dehydration technique for preparing clay samples for electron microscopy.

Critical Point Drying

Critical point drying was not used extensively in studies of clay fabric, but its importance is becoming recognized as a necessary dehydration method in order to maintain the particle-to-particle integrity of a clay sample. Smart and Tovey (1982) summarized several techniques for drying and preparing samples depending upon soil types and natural conditions of the samples. The critical point method was used extensively for biological applications in order to preserve delicate membrane tissues for electron microscopy (Hayat and Zirkin 1973). It was introduced by Kistler (1932) to allow the preparation of delicate mineral gels, such as silica gel in which air is the fluid phase. The technique involves a series replacement from water to alcohol or liquid CO_2 and it has been described thoroughly for biological specimens by Anderson (1951) and Cohen et al. (1968). Gillott (1969) applied the technique successfully to clay sediment, using both alcohol and CO_2 for critical point drying. Gillott claimed that the appearance of the fabric was identical in both cases. Liquid CO_2 has a lower temperature at the critical point (31°C at 7391 kPa) than does alcohol (243°C at 6525 kPa), so CO_2 is more convenient for routine laboratory use. Other fluids also have been used for critical point drying (Cohen et al. 1968). Unfortunately, water does not possess favorable critical characteristics (374°C at 9542 kPa), which makes it impractical for most laboratory applications.

The critical point technique is a marked improvement over other dehydration methods, because surface tension forces are avoided. At the critical temperature and pressure of a liquid, no boundary exists between the liquid and gas phases; and when the temperature is held above the critical point, the gas may be released until atmospheric pressure is reached. Thus the sample can be dried without experiencing surface tension effects, which are extremely deleterious to clay fabrics. Intermediate fluids such as acetone, alcohol, and amyl acetate are used often as transition media, as it is usually necessary to interpose one of these liquids between the aqueous phase of the sample and the liquid CO_2 or other critical point fluid. Alcohol rather than acetone appears to be preferred, because it produces less severe effects on the clays (U. G. Whitehouse, personal communication 1972).

Bennett (1976) used the critical point drying technique, followed by sample impregnation with very low viscosity epoxy resin (SPURR) under vacuum. Using this technique, samples with very complex fabric, large original water content, and small shear strength retained their particle-to-particle structural integrity.

An interesting technique was developed recently for dehydrating submarine sediment while the sample remains at a hydrostatic pressure equivalent to in situ pressure (Chiou 1980). Samples are recovered using a pressure core sampler that was developed at Texas A&M University (W. R. Bryant, personal communication 1978). Sediment samples are then returned to the shore-based laboratory, where subsampling for clay fabric is carried out in a hyperbaric chamber. The device developed by Chiou (1980) maintains an equivalent in situ pressure on the fabric subsample after its removal from the hyperbaric chamber. The miscible transition fluid, ethyl alcohol, is introduced to the sample under high pressure. When the interstitial water is replaced by alcohol (following testing techniques described by Bennett et al. 1977), the sample is returned to the hyperbaric chambers for placement in amyl acetate, and it is then placed in a critical point drying apparatus. After replacement of amyl acetate with liquid CO_2, the sample is dried above the critical point.

These techniques were developed due to the degradation of core samples that occurs commonly in gassy deltaic sediments upon removal of in situ pressure. Using the techniques described, an effort is made to maintain microstructural integrity of the submarine sediment samples (W. R. Bryant, personal communication 1981). Using semiquantitative techniques and statistical analyses for determining clay particle arrangements, Chiou (personal communication 1981) reported significant differences between the fabric of pressurized samples and the fabric of samples recovered by conventional coring techniques. His techniques offer an interesting new approach to studying gassy submarine sediments, and they will contribute significantly to the study of partially saturated, gas-laden sediments commonly found in shallow marine environments.

Determination of Particle Orientation

A problem that faces investigators of clay fabric and that was solved only partially is the determination of the absolute and the relative orientations of particles in sediments and flocs. There are many qualitative descriptions of particle orientations in the literature, but quantitative descriptions are rare for clay sediments. If absolute orientations (with respect, e.g., to the horizontal) are to be determined, it is mandatory to keep track of the reference position throughout the entire process of obtaining, preparing, and recording the sample.

Semiquantitative estimates of the extent to which clay particles are parallel can, at times, be made from the intensity of X-ray diffraction peaks for the minerals. In a somewhat laborious technique applied to TEM micrographs, the writers determined the angles that particle edges make with a common reference plane in two samples cut at right angles to each other. Smart and Tovey (1982) discuss statistical techniques and various methods of analyzing photomicrographs. They include a discussion of current trends in computer applications that may ultimately be the most effective means of rapid analysis.

Present State of Knowledge

Significant advances have been made in understanding clay microstructure since the advent of electron microscopy, X-ray diffraction, electron diffraction, microprobe analysis, and a wide variety of geochemical and electrochemical tools. Direct observations and quantitative analytical methods are providing superior concepts and models of clay microstructure. These advances portend major breakthroughs in understanding clay fabric and the physico-chemistry of fine-grained sediment during the next decade. The significance of future advances will be realized by applying observational evidence and quantitative analyses of sediment microstructure to a wide range of basic and applied research problems.

Research Problems

Some examples of these important research problems that directly tie the clay microstructure to sediment properties include (1) fabric and fundamental geotechnical properties (water content, porosity, bulk density), (2) fabric and physical property changes as a function of overburden pressure and depth of burial, (3) sediment pore fluid migration and permeability as a function of fabric, (4) microstructure and fabric "signatures" in relation to depositional environment, (5) stress-strain behavior of soils in relation to microstructure, (6) acoustic properties (compressional and shear wave behavior) as a function of fabric characteristics, (7) acoustic impedance mismatch at sediment interfaces as a function of fabric characteristics, and (8) active and passive geochemical properties and processes as a function of fine-grained sediment microstructure.

Observational Evidence

Current concepts of clay microstructure that will be the cornerstones of significant advances to solve the problems listed and many others in the earth sciences are summarized below:

1. Clay particles, either as single or compound units, are the building blocks of clay fabric. These particles form complex microstructures that create pores of various geometries. These pores contain either liquids or gases, or both. Sedimentary pores of variable size, shape, and geometry are passive characteristics, and they are a direct function of the particle-to-particle arrangements and fundamental particle sizes.
2. Different clay mineral species respond uniquely to both the chemical environment and mechanical forces.
3. Natural sedimentary deposits and fine-grained soils are composed generally of complex microstructural units of many sizes, shapes, and forms.
4. Microstructural units are composed not only of single platelike particles but also of compound particles held together by physico-chemical forces. Compound particles may include domains, aggregates, floccules, clusters, crumbs, organic compounds, and in some cases, cements.
5. Freshly deposited sediment may consist of multiplate particles, such as domains.
6. Subtle and presently poorly defined relations exist between fabric type and microstructure and the mode of deposition and depositional environment.
7. The chemical environment plays a dominant role during the initial stages of fabric formation. Mechanical forces predominate in producing changes of fabric in the postdepositional stages of sediment diagenesis.
8. Biological activity may play a significant role in the formation of initial fabric units and in the alteration of fabric in sedimentary deposits.
9. Dynamic processes acting on or within a sedimentary deposit can alter the fabric both temporally and spatially.

Further discussion of the present state of knowledge and the significance of current concepts of clay microstructure is presented in the remainder of this text.

Fundamental Characteristics and Models

Clay fabric studies of selected Mississippi Delta and Deep Sea Drilling Project (DSDP) submarine sediments (Bennett 1976; Bennett et al. 1977; Bowles 1968a; Mathewson et al. 1972) have revealed a few features of clay fabric that appear to be diagnostic of material composed largely of smectite and illite. On the basis of these studies, tentative fabric models were developed by Bennett (1976), which integrated basic characteristics of the clay fabric (see fig. 3.37). A few fundamentals of sediment clay fabric were established:

1. The basic building blocks of clay fabric are domains or single platelike particles, or both.
2. The basic particles can form flocs or chains. Unlike SEM representations, TEM photomicrographs represent two-dimensional features of the clay fabrics. Therefore, some chains observed by TEM may actually be composed of numerous face-to-face particles oriented in approximately the third dimension of the field of view (a shingle-type particle chain).
3. The framework of clay fabric can be developed predominately by domains, single platelike particles, flocs, chains, or a combination of these basic fabric entities.
4. High-porosity submarine sediment showing microstructural integrity, which is indicated by the presence of significant natural shear strength, requires well-established particle-to-particle contacts.
5. In order to build a high-porosity sediment that possesses microstructural integrity, particle-to-particle contacts must be maintained commensurate with an increase in the void ratio (volume of the voids to the volume of the solids).
6. The most efficient method of building a high-porosity sediment is through the development of long chains (edge-to-edge or stepped face-to-face particle contacts, or a combination of these features), generally in random arrangement. Very high porosity sediment is achieved by combining these features with a decrease in particle thickness. Particle thickness is greater in sediments composed of domains rather than single platelike particles.
7. Very low porosity sediment can be achieved efficiently by face-to-face particle packing, resulting in a large degree of preferred particle orientation.

These characteristics of clay fabric form a basis for the development of various models considered to be diagnostic of submarine sediment composed predominantly of smectite and illite. Submarine sediment composed chiefly of other types of clay minerals may develop somewhat different particle-to-particle clay fabric relations, but the fundamental relations should hold for most clay sediments. Further studies of sediment clay fabric may require some modification or addition to the proposed fabric models (Bennett 1976). Based on the observations made during that study and through the examination of the fabric presented by Bowles (1968a), approximate tentative values for void ratios characteristic of the various fabric models have been suggested (fig. 3.37).

The models demonstrate that the void ratio can be increased efficiently simply by increasing the lengths of the chains in generally random orientation. Low void ratios, conversely, are characterized by essentially no chains but by a relatively high density packing of clay particles with a tendency towards preferential particle orientation. Clay sediments possessing very low void ratios can be developed efficiently by close particle-to-particle packing (development of sheets) with strong preferred overall orientation of particles (fig. 3.37).

The proposed models of sediment clay fabric can be used as guides to further investigations of sediments, particularly studies focused on various depositional environments. These models should not be considered as characteristic of all submarine clays until they can be verified for other submarine clay types. They are, however, useful in clarifying the particle-to-particle characteristics related to selected mass physical properties, such as void ratio, porosity, wet bulk density, sensitivity, shear strength, and acoustical properties. In addition, the clay fabric representations establish a framework for further studies that will relate diagnostic features of the fabric (particle-to-particle associations) to the behavior of clay sediments under static and dynamic loads.

In reference to various fabric models proposed in the literature, an important feature characteristic of the Mississippi Delta clays investigated by Bennett (1976) and Bennett et al. (1977) is the presence of domains as the basic framework of the fabric. The fabric characteristic of these clay sediments appeared to agree reasonably well with the model of particle arrangement in the flocculated clay proposed by Moon (1972), with the following exceptions (fig. 3.26):

1. The Mississippi Delta clays are characterized by a wide range of clay particle size that is not depicted in the model proposed by Moon. Thus the natural fabric is more varied and complex.
2. Moon proposed a considerably smaller degree of preferred particle orientation for consolidated clay than was observed in the low-porosity Mississippi Delta clays.
3. The delta low-porosity, consolidated clays revealed the thin, long voids associated with the large degree of preferred particle orientation not shown by Moon.

It is also noteworthy that the clay fabric of the relatively high porosity sediment rather closely resembled the turbostratic arrangement proposed by Aylmore and Quirk (1960, fig. 3.11). Meade (1964) questioned strongly whether this type of fabric was a significant component of natural sediments. Meade (1964) stated that little or no evidence supported the formation of either oriented fabrics, preferred, or turbostratic arrangements during natural consolidation. Bowles et al. (1969) found that the clay fabric of most undisturbed submarine sediments from selected areas of the Gulf of Mexico was characterized by a loose, open, random arrangement of particles. They suggested that clay particles observed in samples subjected to small normal stresses were packets of particles oriented in parallel that may have been an original feature of the fabric prior to loading. The evidence was not conclusive but certainly suggestive of the presence of domains. Although features similar to the Mississippi Delta fabrics were observed by Bowles et al. (1969) in other Gulf of Mexico sediments, observations clearly point out significant differences that can account for the variations in certain measured mass physical properties. An example of this is the relation between fabric and void ratio (see fig. 3.34). There was a striking difference between the clay fabric of the deep-sea pelagic clay sample from the Pacific Basin and the Mississippi Delta fabric; again, selected mass physical properties were related to the observed fabric (Bennett 1976). Recently Guorui (1983) attempted to relate the fabric characteristics of a few marine soils to their geotechnical properties. Guorui developed a general descriptive classification scheme for clay microfabric, in which fabric features are associated with soil engineering characteristics, such as strength, porosity, compressibility, sensitivity, and rheology.

5 *Clay Fabric and Environment*

The clay microstructure of flocculated, suspended sediments and of sedimentary deposits depends on a long series of geologic and/or oceanographic processes and events, and on the types of source material available, and on the environmental setting. These processes and factors produce primary and postdepositional sediment properties in time and space. The primary depositional properties include textures, mass physical and mechanical properties, and soil state (fluid-plastic-solid), and they are determined by particle size, mineralogy, particle size distribution, and microstructure. As such, the microstructure is a critical element in determining the fundamental properties of aeolian, fluvial, lacustrine, terrestrial, and submarine deposits. Sedimentary particles, detrital and biogenic, interact in response to mechanical and electrical forces during transport to depocenters. These forces are important in determining the fabric of the transported material and the freshly deposited sedimentary material. Deposited particles establish the primary properties of the sediment. Once deposited, postdepositional processes come into play. Postdepositional processes and effects can be grouped into three major categories—biogeochemical, physical-mechanical, and mechanical (fig. 5.1, Bennett and Nelsen 1983).

Biogeochemical processes and effects include but are not limited to biological activity (bioturbation), gas production, intrastratal geochemical changes (salinity gradients), and authigenic mineralization. Most of these factors can produce changes in porosity and water content by alteration of the microstructure and therefore result in significant changes in the geotechnical properties, including stress-strain behavior and strength.

Physical-mechanical processes are differentiated here from purely mechanical processes by being generally shorter term events rather than processes that occur over geological time spans. Physical-mechanical processes include surface and internal wave forces and their interaction with bottom sediments, bottom currents that place stresses on surficial deposits, ice gouging and loading, and short-term seismic activity, which may cause collapse of the fabric and loss of strength of the sedimentary deposit.

The mechanical processes and effects are generally geologically long term. Sediment dewatering and consolidation are important mechanical processes that result in lowering porosities, water contents, and permeabilities but in increasing wet bulk density. Increasing shear strength with depth of burial is the rule in normally

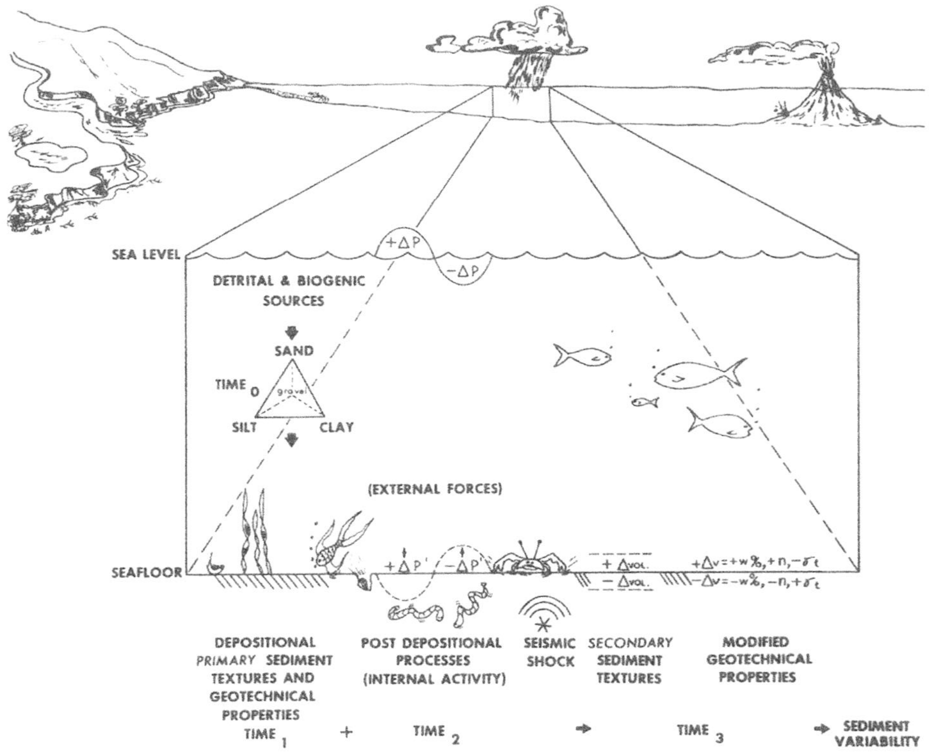

Figure 5.1 *Important processes that determine the type and nature of the source material and affect the primary and postdepositional fabric and geotechnical properties of sediment (modified from Bennett and Nelsen 1983).*

consolidated material. Sedimentation produces overburden stresses that can differ by orders of magnitude, depending upon the depositional environment and volume of material. Creep is another long-term geological process that may result in postdepositional changes in the sediment fabric.

The processes mentioned modify the primary depositional properties and characteristics of the sediments and produce the postdepositional conditions. These vary significantly both regionally and locally.

Environmental Effects on the Development of Clay Fabric

The relative degree to which various factors interact and affect the ultimate nature of the clay fabric particle-to-particle relations, particularly during the initial stages of development, is poorly known. However, the obvious environmental factors that play a role in the formation and development of the clay fabrics in marine environments, and that have been recognized in one case (the Mississippi Delta; Bennett 1976; Bennett et al. 1977) are summarized in the following sections.

Initial Stage

Clay particles experience a transition in the chemical environment as sediment is carried from riverine fresh waters to a saline marine environment. Given sufficient suspended sediment concentration, flocculation or some form of clay particle aggregation would be expected on contact with sea water. Whitehouse et al. (1960) have reported the minimum concentrations of major cations (sodium, magnesium, calcium, and potassium) required to initiate flocculation of montmorillonite, illite, chlorite, and kaolinite. Montmorillonite required the greatest concentration of cations to cause flocculation. Concentrations of sodium and magnesium equivalent to approximately 20% of open ocean water salinities caused flocculation of montmorillonite, whereas calcium and potassium caused flocculation at concentrations equivalent to approximately normal seawater salinities. Bouma et al. (1971) reported relatively high salinities near the channel mouths and over extensive areas of the Mississippi prodelta environment despite the vast influx of fresh water through the distributary channels of the delta. In many other riverine areas, a stable, tapering wedge of saline water lying beneath the fresher river water extends well up the river from the coastline.

The organic and inorganic compounds carried by rivers can be expected to alter the development of clay fabric significantly, but these effects in other than simple laboratory preparations are virtually unstudied. For instance, various phosphates present in river waters may act as peptizers, which tend to disperse the fine particulate clays, whereas other chemical compounds may tend to cause flocculation. As mentioned earlier, certain organic materials of undetermined composition affect the clay mineral surface charges, allowing them to approach one another more closely. These factors, undoubtedly, play a significant role in the initial formation of clay fabric. However, the degree to which these and other chemical aspects influence the particle-to-particle interaction in natural systems is not known.

Turbulence, particularly in nearshore and riverine environments, certainly plays a role in the development of clay fabric. Turbulence as a form of mechanical energy would tend to disperse and disaggregate loosely flocculated particles in the water column. Currents and turbulence would tend to modify or disturb freshly deposited clays, and when they are of sufficient strength to generate stress levels large enough to overcome the energy of particle attraction, they could cause resuspension. Conversely, this form of mechanical energy could provide the force required to bring clay particles sufficiently close together and over a local potential energy maximum in order to allow particle aggregation (see fig. 2.13, curve d). Kranck (1980) discussed experiments on the significance of flocculation in the settling of fine-grained sediment in still water. She reported that turbulence increases interparticle collision, which encourages flocculation, but it also restricts flocculation by disrupting flocs, particularly larger ones. Kranck interpreted preliminary experiments on flocculation in turbulent water as indicating that similar textures were produced under turbulent and still conditions, and both were similar to those found in nature. She concluded that this similarity indicates that flocculation is important in the deposition of most clay and fine silt sediments and of many coarser sediments. Obviously, all these factors increase the complexity and control the ultimate nature and character of the clay fabric. Studies that address these questions and processes as related to clay fabric and the physico-chemistry need to be completed.

Postdepositional Stage

Undoubtedly, benthic biological processes in freshly deposited sediment influence the nature and characteristics of clay fabric, particularly in areas of great faunal activity. The influence of biological activity on the mass physical properties of submarine sediment has been studied to a limited extent (Moore 1931; Piper and Marshall 1969; Rhoads and Young 1970). Nevertheless, little is known about the changes in clay fabric and the physico-chemistry that results from biological processes and activity. Bioturbation probably influences freshly deposited clay sediments, but this may be relatively insignificant in some delta environments due to the rapid rates of sediment deposition that usually inhibit the degree of biological activity.

Postdepositional cementation probably influences clay particle rearrangement during dewatering and natural consolidation of fine-grained sediment. Nacci et al. (1974) have alluded to this factor, but much more research is required to assess the importance of the chemical cementation processes in relation to their influence on clay fabric. Evidence of cementation was not observed in the clay fabrics of the Mississippi Delta clays, the DSDP clay sample (Bennett 1976), or the Gulf of Mexico clays studied by Bowles (1968a). Apparently, other factors, still undefined, are also important in influencing the force transmission and particle-to-particle interaction during natural consolidation as a function of overburden stress.

Obviously, the presence of excess pore water pressures in sedimentary deposits plays a significant role in determining the degree of consolidation and effective stress as a function of depth of burial (Bennett et al. 1982). Large rates of fine-grained sediment deposition coupled with very low permeabilities can result in the occurrence of excess pore pressures. The clay fabric and the particle-to-particle arrangements determine the sizes of the pore spaces and their geometry, thus ultimately controlling pore water flow (permeability).

One industry that has a critical interest in the permeability of deposits is the petroleum industry (Hower et al. 1980). Reservoir deposits with low permeability make recovery of oil and gas difficult and costly. Sediments possessing fabric characteristics that produce continuous pores and permit large flow rates are attractive for petroleum recovery and production. The permeability of deposits is altered by various treatments that change the microstructure of the clays present or that emplace clays of selected microstructure. The disposal of waste clays in the petroleum industry (and in the extractive mineral industry in general) often makes it desirable to control the microstructure of the wastes (see Stanley et al. 1984 and references therein).

Agriculture is another industry that has long understood its dependence on soil characteristics, including soil microstructure (see, e.g., Black 1968 and Brady 1974). The properties commonly viewed as being critical have included permeability (control of access of water, soluble nutrients, and gases to the roots), fabric (water infiltration, evaporation, fluid capacity, root-soil contact area, and shrinkage-swelling properties), and strength (seedling emergence, root development, root penetration, mechanical support of plants, erodibility, and soil compaction). Investigations of these problems continue at an intense pace (see, e.g., Carlan et al. 1985; Gerik and Morrison 1985; Wilson et al. 1985). With the start of the biotechnology revolution during 1980, the importance of soil microstructure in agriculture began to be appreciated more fully. Effects of soil microstructure on the community of organisms that colonizes soils are particularly important. Minor changes in this community can alter crop productivity dramatically (Lynch 1983).

Depth of Burial Versus Fissility

To a geologist, *compaction* refers to a decrease in the volume of sediment as a result of compressive stress. Commonly, this compressive stress results from the weight of sediments (the overburden) deposited later. Compaction is essentially equivalent to *consolidation* as used by the soil engineer with reference to a gradual decrease in water content and decrease in total sample volume as a function of an imposed load (usually a vertical load) on a sediment or soil sample. Perhaps the earliest serious study of clay compaction was made by Sorby (1908), who recognized that the compaction of sediment results primarily from a change in porosity. He carried out both field and laboratory investigations of sediment porosity. Numerous subsequent studies have been concerned with deciphering relations among porosity changes, depth of burial, overburden stress, the presence and degree of fissility in certain shales, stages of compaction, and relation with organic materials and depositional environment (Hedberg 1936; Weller 1959; White 1961; Meade 1961, 1966; Gipson 1965, 1966; Odom 1967; Gillott 1969; Rieke and Chilingarian 1974). Although the details of these studies are not explored fully here, a few of the more important aspects are summarized.

Hedberg (1936) recognized four integradational stages of compaction of a fluid mud to a dense shale or slate, and he related these stages to changes in porosity as a function of compressive stress. He recognized the importance of the mechanical rearrangement of particles during the early stages of compaction. Weller (1959) was concerned with a quantitative approach to the evaluation of overburden stress with depth of burial, and he developed numerous data curves that relate these factors to porosity. Meade (1966) pointed out the important inverse relationship between particle size and porosity, particularly within the range of stress of up to about 10 MPa. He recognized that the effect of particle size may be strong enough to obscure the expected decrease in porosity with increasing depth of burial. The significance of this observation should be emphasized. When mud and sand are subjected to an equivalent depth of burial or overburden stress, the porosity of the mud commonly is observed to be greater. This observation is contrary to intuition because the mud may have a smaller solid mass per unit volume than the sand and therefore might be expected to occupy a smaller volume. However, the fabric of the material determines the relation of voids to solids, thus the amount of pore space available for fluid. Sedimentary materials composed of very fine-grained particles are characterized commonly by complex fabrics that produce many small voids (high void ratio), relatively low permeabilities, and high porosities. Thus compaction is inhibited significantly in fine-grained sedimentary deposits.

White (1961) and Gipson (1965, 1966) found that a greater degree of clay mineral preferred particle orientation was associated with fissile shales, and random orientation was characteristic of the more massive argillaceous rocks. Similar conclusions were drawn by Gillott (1969) during a study of the fabric of argillaceous rocks and soils. Likewise, Odom (1967) found a strong correlation between the degree of clay mineral preferred particle orientation and fissility, and the presence of organic material. Massive shales showed essentially random orientation of clay particles. Direct correlation of fissility and organic material was reported also by Gipson (1965). Numerous investigators speculated that the presence of organic material tends effectively to reduce the net negative charge of clay particles, thus allowing them to move close together more efficiently in a face-to-face arrangement. Both Gipson (1966) and Odom (1967) found no correlation between depth of burial and preferred clay mineral orientation.

Although considerable conjecture exists concerning changes in porosity with increasing depth of burial and development of fissility in shale, the most conclusive evidence to date indicates that preferred particle orientation is related to fissility, but the degree of preferred clay mineral orientation is not related directly, in most cases, to depth of burial. The general decrease in porosity with increasing depth of burial and overburden stress varies considerably from sediment to sediment.

Evidence pointed out and alluded to by many investigators studying clay fabric, compaction (consolidation) of muds, and the fabric of shales has indicated the importance in understanding the geochemical conditions, environments of deposition, and related geological factors that set the stage for the initial formation of clay fabric and the fabric characteristics at the time of deposition. Apparently, the fabric at the time of deposition plays an important role in determining the ultimate response of the clay particles to overburden stress and the degree to which consolidation will ensue. There is a need for detailed studies that focus on the delineation of geochemical conditions that influence the formation of clay floccules in the water column, the flocculation or sustained dispersion of clay particles from fresh water to saline water under varying natural geochemical conditions (such as the influence of organics and peptizers on flocculation and dispersion). Very little is known of the influence of these conditions on the development of clay fabric in the natural environment. Virtually nothing has been published concerning the effects of biological activity on the depositional and postdepositional clay fabric. It is not known how the environmental factors such as currents, waves, turbulence, and other kinds of mechanical energy influence the development of clay fabric in the water column and at the sediment-water interface. Essentially all previous studies have been concerned with postdepositional sediment and rock fabrics.

The study of Mississippi Delta sediments (Bennett 1976) was confined to the delineation of postdepositional clay fabrics of selected cores from the prodelta environment. This study recognized many of the limitations discussed in the preceding paragraph, but found a large degree of preferred clay particle orientation of sediment buried to depths of not more than 120.5 m below the mudline (corresponding to an overburden stress of not greater than 895 kPa). These clays contrast sharply with the overlying high-porosity sediments which show essentially no strong preferred particle orientation to depths of at least 37 m below the mudline (corresponding to an overburden stress of not greater than 202 kPa). These delta sediments with preferentially oriented clay particles are obviously advanced in the development of fissility (attaining nearly the degree of fissility observed commonly in shales) despite their rather limited depth of burial and corresponding relatively small overburden stress. These sediments contrast sharply with the DSDP deep-sea submarine clay sample, which has a very high void ratio (3.5) and a high porosity (78% , even though it was sampled from a depth of 144 m below the mudline. Clearly, factors other than increasing burial and corresponding overburden stress influence the degree of preferred clay particle orientation, the development of fissility, and the degree of compaction (consolidation) of a particular submarine sedimentary deposit.

Although a few typical fabrics have been classified (honeycomb, cardhouse, turbostratic, bookhouse, and stair-stepped) recent studies have demonstrated that domain-type particles rather than single-plate particles are the rule for most natural sedimentary clayey materials. The unlimited variation in particle size, shape, and composition, coupled with depostional environment, physicochemistry and transportation of sedimentary detritus through changing environmental conditions, all increase the complexity and control the ultimate nature of the sediment fabrics of a particular deposit. Future investigations designed to clarify the nature of clay fabric in relation to sediment mass physical properties and the behavior of clays in relation to dynamic and physico-chemical processes must take these facts into account.

The Principle of Chemical Irreversibility of Clay Fabric

An important generalization that can be derived from studies of clay fabric and particularly from some investigations of techniques for sample preparation (Bennett 1976; Bennett et al. 1977) is the *principle of chemical irreversibility*. Present evidence indicates that, in general, the attractive forces between clay particles in mutual contact in sediments are strong enough and sufficiently independent of the chemical environment to hold the sediment mass together regardless of subsequent changes in chemical conditions. This is particularly evident with regard to further chemical changes that may ensue during natural processes. The significance of this principle of chemical irreversibility of clay fabric for natural sediments is twofold:

1. The chemical environment of clay particles is critical during the initial stages of fabric formation in the water column and at the sediment-water interface.
2. After the initial flocculation of particles and deposition, the chemistry plays a very passive role in the postdepositional changes in the clay fabric and the postflocculated states. Mechanical energy rather than chemical energy plays the dominant role in postflocculation and postdepositional states of clay fabric development.

In order to tie more closely the importance of the physicochemical aspects of clay microstructure to concepts of clay fabric, the concept of the chemical irreversibility of clay fabric as presented by Bennett (1976) and others (Bennett et al. 1977) is reviewed below.

Chemical Environment and Clay Fabric

In previous sections, the authors discussed the importance of the physico-chemistry and chemical environment in general on the development of clay fabric formation. As pointed out earlier, the chemical nature of the fluid medium of both fresh and saline water affects strongly the electrostatic interactions among suspended clay particles, thus determining the fabric characteristics of flocculation or dispersion prior to sediment deposition. During natural processes of weathering, transport, and subsequent deposition of detrital particles, an increasing electrolytic chemical environment is the rule. This is significant particularly for terrestrial sediments that are transported from riverine environments to the marine environment, that is, for a substantial proportion of weathered sedimentary particles.

Considerable evidence indicates natural changes in the marine chemical environment will not cause disaggregation of clay particles that flocculate and clay particles that mutually contact at the sediment-water interface. Whitehouse et al. (1960) alluded to the chemically irreversible nature of marine-flocculated clay particles. These researchers stated that clay materials exposed to marine waters exhibit a dispersal resistance to ultimate deflocculation that is quite different from the resistance to dispersal of similar material not so exposed.

Convincing evidence indicating the chemically irreversible nature of marine flocculated clays was revealed during the analysis of the clay fabric techniques employed in studying the Mississippi Delta and DSDP samples (Bennett 1976). These sediment samples were prepared for TEM clay fabric analysis through the replacement of the interstitial water with a series of miscible fluids for critical point drying. The dry samples were then embedded with SPURR low-viscosity resin. Ethyl alcohol replaces the saline interstitial water, amyl acetate replaces the alcohol, and liquid CO_2 replaces the amyl acetate for critical point drying. As an example, an important factor of the chemical changes that the interstitial fluids undergo is the virtually complete stripping away of the ambient saline interstitial water with large electrolyte content. In addition, the original interstitial saline water possessing a bulk dielectric constant of approximately 78–80 is replaced by fluids characterized by considerably smaller dielectric constants of approximately 24 (ethyl alcohol), 5 (amyl acetate), and 1.6 (liquid CO_2). This type of treatment is rather vigorous in terms of the chemical changes experienced by the interstitial fluids and surrounding chemical atmosphere of the clay particles in a sediment, and it is much more rigorous than would be experienced by marine suspended sediment or submarine sedimentary deposits in nature.

Despite the rather vigorous chemical changes and subsequent embedding of the dry samples with resin, the clay fabric samples retained their particle-to-particle integrity, indicating the existence of attractive forces between clay particles that were strong enough to prevent microstructural degradation (Bennett 1976). This suggests that chemical changes in the marine environment following the flocculation and sedimentation of particles will not alter significantly the fabric of sedimentary material. In addition, subsequent changes in the original character of the fabric require some form of mechanical energy, such as turbulence (in the water column), shearing of sediment or flocs, overburden stress, or biological activity (bioturbation).

The validity of the principle of chemical irreversibility has been denied by Moon and Hurst (1984). They speculate that clay flocs are dispersed by chemical action when they enter anoxic environments. Thus they suggest that one important cause of preferred particle orientation in natural sediments is the accumulation of single particles disjoined from flocs by such chemical activity. Although conditions in anoxic natural waters may tend to minimize flocculation of particles that are already dispersed, the speculation of Moon and Hurst that such conditions cause flocculated particles to disperse is contrary to a large body of evidence. Many saturated sediments are anoxic but still contain flocculated particles. The surficial samples of Mississippi Delta sediment studied by Bennett (1976), for example, were from an anoxic, sulfate-reducing environment and showed no strong preferred particle orientation.

Qualitative consideration of the potential energies of interaction between clay particles can help answer questions about alterations in the fabric of a sediment mass upon changes in its chemical environment. For real clay particles, the relation between the energy of interaction and the distance of separation is appreciably more complex than presented in figures 2.13, 2.14, and 2.15. Derivation of the relation presented in these figures was made possible by incorporating simplifying assumptions, the most restrictive being that the particles are large, identical planes that are always parallel. Real particles are three-dimensional and may have any orientation. Further, the different faces of real particles may be significantly different in terms of their electrostatic potentials; the potentials may even be opposite in sign (Welzen et al. 1981). Thus instead of a simple two-dimensional curve showing the relation between separation distance and potential energy for two particles, a four-dimensional representation would be required. The potential at each point is determined by the effects of all particles in the vicinity, and all are free to move relative to one another. Therefore, theoretical treatment of real systems of clay particles becomes formidable indeed. Nevertheless, qualitative consideration of the properties of a simplified potential surface can give some insight into the alterations of clay fabric. Consider a clay particle in a sediment composed of flat particles. Assume that this particle is free to translate in a plane, while all other particles of the sediment remain fixed. The potential energy as a function of the position of the particle may be represented as in figure 5.2; the position is given by a location in the x, y plane, and the potential energy is given by the vertical contour (producing a graph analogous to an ordinary topographic contour map, but one of very rugged terrain). This

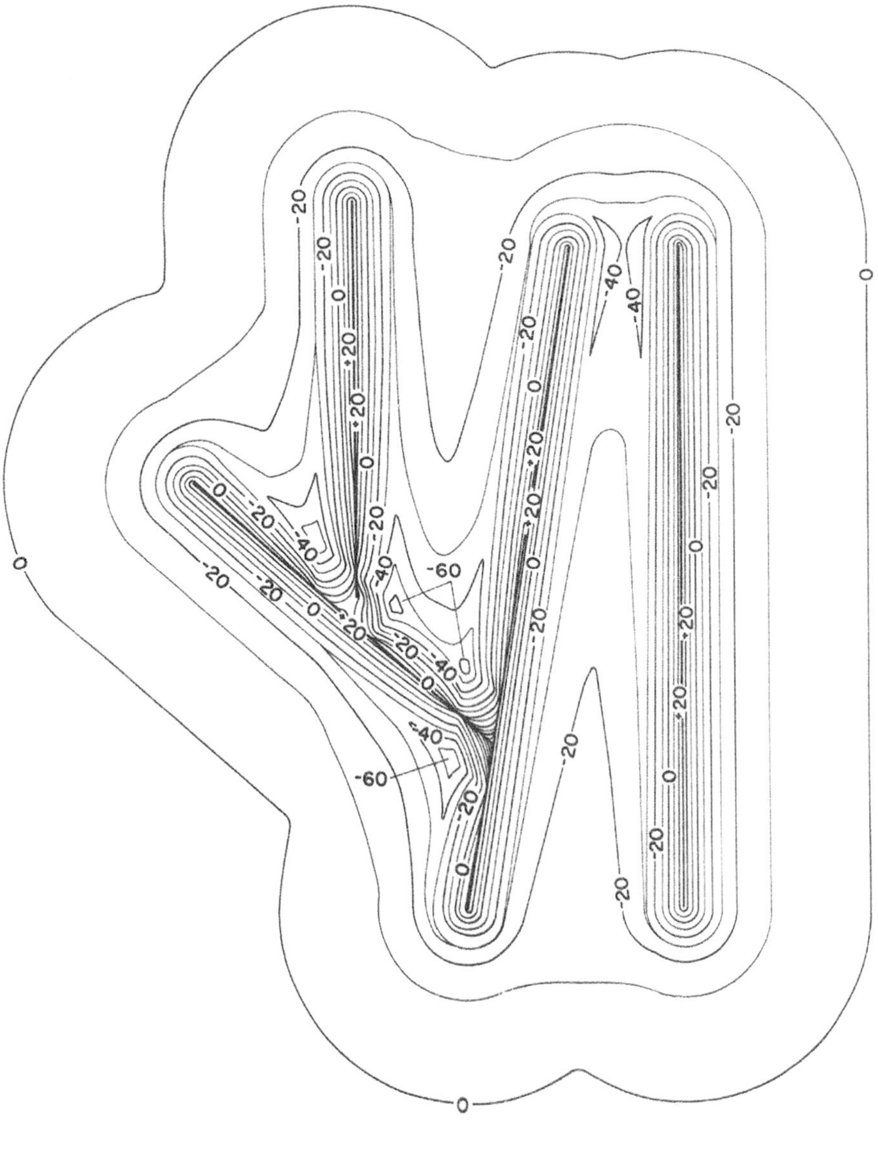

Figure 5.2 *Potential energy in the vicinity of clay particles as felt by another particle that is free to translate in a plane.*

schematic of the potential energy surface was constructed by superposition of the potentials for each of the fixed particles (see curve d, fig. 2.13). The customary sign of the potential energy is used (namely, the mobile particle possesses less potential energy the less positive the contour value), and contour magnitudes are on an arbitrary scale.

Figure 5.3 represents a similar system with a greater electrolyte concentration. In each of the systems, the thermal energy of the particle is small with respect to the contour unit; thus the particle tends to remain approximately in a fixed position at one of the local minima in potential. This is true even though much deeper local minima may be nearby, because the intervening potentials are large. As the system is disturbed more and more energetically, the particle is able to transcend the intervening potential barriers and to move spontaneously to deeper local minima. Therefore, although the fabric formed initially in a sediment may not place particles in their lowest state of potential energy, and the energetics of the entire system may be somewhat altered by change in the chemical environment, appreciable energy must be supplied before the particles can assume a more stable fabric.

Significance of the Principle

The significance of the principle of clay fabric irreversibility for sediment is that the chemical environment surrounding the clay particles is critical during the initial stages of fabric development in the water column and at the sediment-water interface. Subsequent to the initial flocculation of particles and deposition, the chemistry of the interstitial water and ambient water chemistry probably play very passive roles in the postdepositional changes in the clay fabric and in the postflocculated states. Obviously, flocculation in the water column would not be a significant process without a sufficient concentration of suspended sediment providing close particle-particle proximity for particle contact. This does not rule out the fact that postdepositional cementation and authigenic mineralization, in some cases, may influence the ultimate response of the clay particles to physical forces. Mechanical energy rather than chemical energy is most important in the postflocculation and postdepositional stages of clay fabric development.

Additional important implications of the principle of chemical irreversibility are as follows:

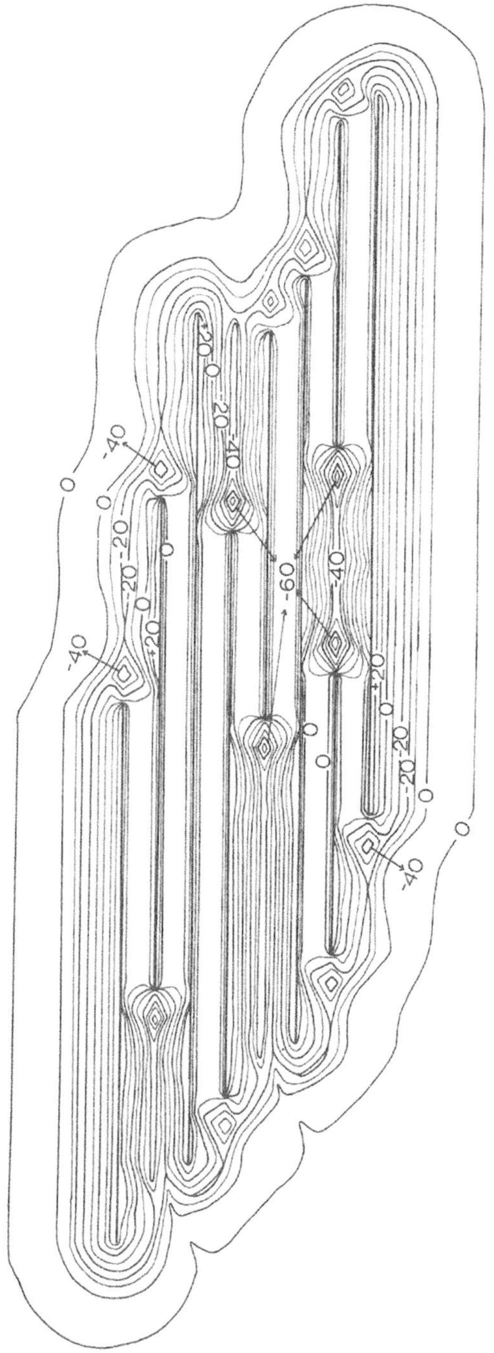

Figure 5.3 Potential energy in the vicinity of clay particles in a system prossessing greater electrolyte concentration than depicted in figure 5.2. Contours represent potential energies as felt by another particle that is free to translate in a plane.

1. Clay fabric samples will maintain structural integrity during careful physical and chemical treatment required for electron microscopy investigation.
2. Unique fabric signatures diagnostic of the chemical environment of deposition are unlikely to be apparent in fine-grained sediment.
3. The relative importance of chemical and mechanical factors in the development of clay fabric can be predicted for particular geological processes.

6 *Summary*

It is appropriate to end this review with a brief statement of the key factors that control clay fabric and of the areas in which a deeper understanding of clay microstructure promises to be fruitful. Direct application of the results of many laboratory studies of clay microstructure to natural sediments and to clayey materials of commercial interest is commonly impeded by the much larger degree of variability in real systems than in laboratory systems. For example, natural systems are usually composed of a variety of minerals, mixed crystals, and amorphous solids. The solids have a variety of sizes and shapes. The chemical composition of the fluid phase is complex, often with traces of a broad variety of compounds of biological origin. As difficult as it is to deal with the complexity of real systems, a greater impediment to a functional understanding of real systems has been the common failure to account adequately for changes with time in conditions that control fabric development. In the following brief discussion, an attempt will be made to identify the crucial factors to be considered in investigating fabric in natural systems.

Natural systems of clayey materials commonly pass through a stage in which the clay-sized particles are in aqueous suspension before the clay deposit forms the fabric of interest. Powerful theoretical tools, based upon physico-chemical principles, have been developed for understanding, even predicting, the fabric of the deposit. It is important to know, however, the state of the material in suspension. Are the clay particles actually present as independent units in suspension, or are they in associations which preexisted in the parent rock? Groups of particles weathered from parent material may remain associated in suspension as foliated stacks or small flocs. In addition to the initial states and types of clays in suspension, the factors that predominate normally in determining the fabric of the material deposited are the sizes of the initial units, the concentrations of the solid particles, the total salt content of the water, and the energy of the environment. The last three factors ordinarily vary during the transit of particles from initial suspension to final position in a sediment.

The principle of chemical irreversibility suggests that further changes in the fabric after initial deposition will probably result from mechanical factors. Application of this principle greatly simplifies the task of understanding how complex, changeable environments produce sediments of a particular clay fabric. To a first approximation, despite the variety of inorganic and organic compounds present in natural systems, the student of clay fabric needs to be concerned primarily with a simple sum of concentrations, the salinity. In many situations, the natural limits on the composition of the aqueous solution minimize differential effects due to the organic material present, the identity of the inorganic ions, the pH, the oxidation-reduction potential, and so on. However, special cases exist and offer significant challenges: In alkali lakes, the ionic strength may be even more important than the salinity. In some circumstances, organic matter may exceed 5% or may have an unusual composition and contain enough surface-active components to affect clay fabric when the salinity is low. It is important to note that the preceding comments apply to fabric studies as first approximations only. Advances in understanding the microstructure of fine-grained sediments along the lines discussed in this review will lead to fabric studies utilizing progressively more sophisticated theoretical constructs as their foundations.

Once the initial clay deposit forms, the mechanical forces provided by its environment are of primary concern. Chemical factors cannot be ignored completely, because they determine the diagenesis of the clay minerals. They provide for the possible precipitation-dissolution of cements or of other solids within the existing fabric. Occasionally they change so radically that the fabric is altered (e.g., during complete desiccation). Important natural sources of mechanical forces in sediments include turbulence due to moving water or to mass movement of sediments, the weight of overlying sediments deposited later, the activities of biological organisms, and acoustic energy from seismic shock.

The scale of fabric observations makes generalization to natural sedimentary deposits extremely difficult. (Fabric samples possess dimensions of cubic micrometers, whereas the deposits they represent may possess dimensions of cubic kilometers.) Perhaps this difficulty is evident most clearly in studies of the effects of sediment biota on clay fabric. For example, casts of many burrowing organisms are so distinctly different from the surrounding sediment that they may be used as markers for specific species. Yet the interior of such a cast may show no significant alteration of fabric. A similar consideration applies for other factors that affect fabric: A very small subsample of material right at a shear plane may be quite different from one a fraction of a millimeter away, or a chemical inhomogeneity may occupy an extremely small volume. In each case, a difficulty occurs, because the sample size on which the fabric is determined is so tiny.

Some of these difficulties with sample size have been overcome by recent studies in which several TEM photomicrographs of a sample were used to form a mosaic. This provides the observer with a continuous, high-magnification view of the fabric over an area larger than is commonly possible using a single photomicrograph taken over a random area of a sample (W.R. Bryant, personal communication 1985).

Some research areas to which a deeper understanding of clay microstructure promises to make important contributions are not difficult to discover, but an exhaustive listing of such areas would be difficult to make. Prediction of the physical properties of clayey masses and of the changes in these properties with changing conditions, such as variations in applied stress, is central to the work of ceramicists and geotechnical researchers. Much progress has been made by both groups through the use of bulk measurements of certain properties, such as water content, but a more fundamental basis for prediction is the microstructure. From understanding the microstructure, one may better predict the stability of natural sedimentary deposits (see work by Pusch 1966, 1970 on quick clays) and have more precise and reliable control of the properties of manufactured articles. Information about fabric may be very useful in understanding the dynamic processes that occur at the interface between sediments and the overlying water. For example, do sedimentary deposits possess fabric signatures that are diagnostic of depositional environment? A more sophisticated understanding of clay microstructure than current knowledge permits would allow researchers to deal quantitatively with fundamental questions regarding these phenomena. These questions include the propagation of acoustic energy through sediments and sedimentary rocks, acoustic energy reflectivity and attenuation, and the degradation or retention of the structural integrity of soft sediments subjected to seismic shock or pressure perturbations caused by wave loading in shallow water.

The present state of knowledge of clay microstructure indicates that the basic properties of rigidity of soft, clayey sediments depend strongly on the fundamental nature of the fabric, the particle contacts, and physico-chemistry. As shown earlier, some of the geotechnical properties of sediments can be related to fabric (Bennett et al. 1977). Rigorous quantitative and statistical treatments of clay fabric would undoubtedly increase our ability to understand and to deal with the complex interrelation between sediment and sedimentary rock behavior and environmental (natural and human-induced) stresses. Other areas in which knowledge of clay microstructure studies may be valuable include diverse environmental problems:

1. Disposal of radioactive wastes in the seabed (relations between clay fabric and permeability and potential migration of pollutants (Bennett et al. 1985); fabric and permeability changes in relation to thermal and chemical effects of the wastes themselves).
2. Removal of particulate matter due to mining operations from lakes and oceans (flocculation, floc disruption, and rates of sedimentation).
3. Control of soil erosion and nutrient depletion in agricultural lands (fabric effects on soil stability and interactions between fabric, cultural practices, and crop productivity).
4. Prediction of geological hazards to human-emplaced objects on the seafloor (fabric determinants of static and dynamic strengths of sediments).

Fabric studies of clayey deposits and shales may provide information on the depositional environment, which in turn may be useful in studying the geologic record. Fabric investigations may provide the basis for predicting important geological processes, such as potential landslides and failures, mass movements of the seafloor, and transport of fine-grained sediment.

Research on clay microstructure may be directed profitably in several areas. A prime area so far only partially investigated is the fabric of materials in suspension and at the interface between suspension and sediment in natural settings. Although much of the practical interest in microstructure is associated with natural sediments, laboratory systems have been more amenable to study. Studies need to be extended much more fully to natural systems, with proper attention paid to the questions of scale, mentioned earlier. An impediment to a fuller understanding of natural systems is the currently unsatisfactory experimental data and theoretical model for the behavior of water in bulk and at interfaces

(Klier and Zettlemoyer 1977; Martini 1981; Oliphant and Low 1983; Mulla et al. 1984). Not all investigators have paid sufficient attention to the multitude of problems of sample preparation for fabric studies. Another area that needs rapid development is the relation of microstructure to bulk properties and geotechnical properties of materials. It would appear that an appropriate approach would parallel that of statistical mechanics: The dispersion of suitable micro-characteristics would be taken into account in order to predict bulk characteristics. A parallel problem is the quantitative description of microcharacteristics (particle size, shape, orientation, and microscale particle-to-particle strength). Quantitative expression of microcharacteristics has been developed only in rudimentary fashion.

The problems of research methodology and experimental design for solving some of the questions listed are formidable. These research tools will be developed piecemeal over several years. They will contribute not only to the field of clay microstructure, but they will aid in answering important questions in other major disciplines of the earth sciences, soil engineering, and agronomy. With the presently available instrumentation, a knowledge of past difficulties, and a solid foundation of current concepts, the time is ripe for pursuing studies concerning early stages of fabric formation and the postdepositional nature and processes of naturally occurring fine-grained mineral sediments.

In future research endeavors, the student of clay microstructure must review virtually all environmental factors in the actual depositional and transitory environments. Physical, mechanical, chemical, and biological conditions and processes may all be important in both natural environments and in the laboratory. Appropriate consideration of their influences on the microstructure and on the experimental results is of paramount importance to the success of research on sediment and sedimentary rock microstructure.

The current state of understanding of clay microstructure is advanced sufficiently that consideration of microstructure can provide insights into a number of scientific and technical problems in the areas of geology, geochemistry, acoustics, petroleum engineering, structural engineering, ceramics, soil science, and agriculture. In these and other disciplines, the developments discussed in this review portend an explosive growth in application of information on clay microstructure.

References

Almagor, E., and G. Belfort, 1978, Relaxation studies of adsorbed water on porous glass: Journal of Colloid and Interface Science, v. 66, p. 146–152.

Andersland, O. B., and A. G. Douglas, 1970, Bonding, effective stresses, and strength of soils: Journal of the Soil Mechanics and Foundation Division, Proceedings of the American Society of Civil Engineers, v. 96, p. 1073–1077.

Anderson, T. G., 1951, Techniques for the preservation of three-dimensional structure in preparing specimens for the electron microscope: Transactions of the New York Academy of Science, v. 13, p. 130–133.

Aylmore, L. A. G., and J. P. Quirk, 1959, Swelling of clay-water systems: Nature, v. 183, p.1752–1753.

—— 1960, Domain or turbostratic structure of clays: Nature, v. 187, p. 1046–1048.

Bancroft, W. D., 1926, Applied colloid chemistry, general theory, 2nd ed.: New York, McGraw-Hill Book Co., 489 p.

Barden, L., 1972, The influence of structure on deformation and failure in clay soils: Geotechnique, v. 22, p. 159–163.

—— 1973, Theme 2, Macro- and microstructure of soils, Proceedings of the International Symposium on Soil Structure, Gothenburg, Sweden: Stockholm, Swedish Geotechnical Society, p. 21–26.

Barden, L., and G. R. Sides, 1970, Engineering behavior and structure of compacted clay: Journal of the Soil Mechanics and Foundation Division, Proceedings of the American Society of Civil Engineers, v. 96, p. 1171–1200.

—— 1971, Sample disturbance in the investigation of clay structure: Geotechnique, v. 21, p. 211–222.

Barden, L., G. R. Sides, and J. P. Karunaratne, 1970, A microscopic examination of aspects of clay structure, Proceedings of the 2nd Southeast Asian Conference on Soil Engineering, Singapore: London, International Society of Soil Mechanics and Foundation Engineering, p. 67–72.

Barelli, N., 1980, Mineralogical, analytical and microstructural studies of some soils used as road foundation (abstr.): 17th Annual Meeting, The Clay Minerals Society, 29th Annual Clay Minerals Conference, Waco, Texas: Waco, Texas, Baylor University, p. 19.

Bennett, R. H., 1976, Clay fabric and geotechnical properties in selected submarine sediment cores from the Mississippi Delta: PhD dissertation, Texas A&M University, College Station, Texas, 269 p.

Bennett, R. H., and W. R. Bryant, 1973, Submarine sediment microstructure (abstr.): 10th Annual Meeting, The Clay Minerals Society, 22nd Annual Clay Minerals Conference, Banff, Canada: Bloomington, Indiana, The Clay Minerals Society, p. 22.

—— 1976, Clay fabric and selected geotechnical properties: Mississippi Delta submarine sediment cores (abstr.): 13th Annual Meeting, The Clay Minerals Society, 25th Annual Clay Minerals Conference, Corvallis, Oregon: Bloomington, Indiana, The Clay Minerals Society, p. 32.

Bennett, R. H., W. R. Bryant, and P. J. Burkett, 1985, Microstructure of red clays from the northwest Pacific deep-sea basin: poster sessions (abstr.): International Clay Conference, Denver, Colorado.

Bennett, R. H., W. R. Bryant, and G. H. Keller, 1977, Clay fabric and geotechnical properties of selected submarine sediment cores from the Mississippi Delta: NOAA Professional Paper 9, 86 p.

—— 1979, Clay fabric and related pore geometry of selected submarine sediment: Mississippi Delta: Scanning Electron Microscopy, v. 1, p. 519–524 and 424.

—— 1981, Clay fabric of selected submarine sediments: Fundamental properties and models: Journal of Sedimentary Petrology, v. 51, p. 217–232.

Bennett, R. H., J. T. Burns, T. L. Clarke, J. R. Faris, E. B. Forde, and A. F. Richards, 1982, Piezometer probes for assessing effective stress and stability in submarine sediments, *in* S. Saxon and J. K. Nieuwenhuis, eds., Marine slides and other mass movements: New York, Plenum, p. 129–161.

Bennett, R. H., D. N. Lambert, M. H. Hulbert, H. R. Schmoll, and B. M. Bohlke, 1980, Clay fabric of sediment from various depositional environments (abstr.): 17th Annual Meeting, The Clay Minerals Society, 29th Annual Clay Minerals Conference, Waco, Texas: Waco, Texas, Baylor University, p. 22.

Bennett, R. H., and T. A. Nelsen, 1983, Seafloor characteristics and dynamics affecting geotechnical properties at shelfbreaks, *in* D. J. Stanley and G. T. Moore, eds., The shelfbreak: Critical interface on continental margins: Tulsa, Oklahoma, SEPM, p. 467.

Bhaskaran, R., 1972, The microstructure of dispersed and flocculated samples of kaolinite, illite, and montmorillonite: Discussion; Canadian Geotechnical Journal, v. 9, p. 233–234.

Biscaye, P. E., 1965, Mineralogy and sedimentation of recent deep-sea clay in the Atlantic Ocean and adjacent seas and oceans: GSA Bulletin, v. 76, p. 803–832.

Bjerrum, L., and I. Rosenqvist, 1956, Some experiments with artificially sedimented clays: Geotechnique, v. 6, p. 124–136.

Black, C. A., 1968, Soil-plant relationships: New York, John Wiley and Sons, 792 p.

Bochko, R., 1973, Types of microstructural elements and microporosity in clays (based on scanning electron microscopy studies), Proceedings of the International Symposium on Soil Structure, Gothenburg, Sweden: Stockholm, Swedish Geotechnical Society, p. 97–101.

Bockris, J. O., and A. K. N. Reddy, 1970, Modern electrochemistry: New York, Plenum, 1432 p.

Bohlke, B. M., and R. H. Bennett, 1980, Mississippi prodelta crusts: A clay fabric and geotechnical analysis: Marine Geotechnology, v. 4, p. 55–82.

Bolt, G. H., 1956, Physico-chemical analysis of the compressibility of pure clays: Geotechnique, v. 6, p. 86–93.

Bouma, A. H., F. B. Chmelik, and R. Rezak, 1971, East Bay, Mississippi River Delta: Gulf Coast Association of Geological Sciences Transactions, v. 21, p. 273–289.

Bowles, F. A., 1968a, Electron microscopy investigation of the microstructure in sediment samples from the Gulf of Mexico: PhD dissertation, Texas A&M University, College Station, Texas, 132 p.

—— 1968b, Microstructure of sediments: Investigation with ultra-thin sections: Science, v. 159, p. 1236–1237.

Bowles, F. A., W. R. Bryant, and C. Wallin, 1969, Microstructure of unconsolidated and consolidated marine sediments: Journal of Sedimentary Petrology, v. 39, p. 1546–1551.

Bradford, S. C., 1918, The theory of gels: Biochemical Journal, v. 12, p. 351–381.

Brady, N. C., 1974, The nature and properties of soils, 8th ed.: New York, Macmillan, 672 p.

Bragg, W., G. Shearer, and J. W. Mellor, 1922, Notes on the crystalline structures of some china clays examined by the X-ray powder method: Transaction of the Ceramic Society, v. 22, p. 106–110.

Brenner, H., and L. J. Gajdos, 1981, London-van der Waals forces and torques exerted on an ellipsoidal particle by a nearby semi-infinite slab: Canadian Journal of Chemistry, v. 59, p. 2004–2018.

Brewer, R., 1964, Fabric and mineral analysis of soils: New York, John Wiley and Sons, 470 p.

Brewer, R., and A. D. Haldane, 1957, Preliminary experiments in the development of clay orientation in soils: Soil Science, v. 84, p. 301–309.

Burnham, C. P., 1970, The micromorphology of argillaceous sediments, particularly calcareous clay and siltstones, *in* D.A. Osmond and P. Bullock, eds., Micromorphological techniques and applications: Agricultural Research Council Soil Survey Technical Monograph, v. 2, p. 83–96.

Byers, C. W., 1974, Shale fissility: Relation to bioturbation: Sedimentology, v. 21, p. 479–484.

Cabrera, J. G., and I. J. Smalley, 1971, Engineering behavior and structure of compacted clay: Discussion and reply: Journal of the Soil Mechanics and Foundation Division, Proceedings of the American Society of Civil Engineers, v. 97, p. 802–805.

Carlan, W. L., H. F. Perkins, and R. A. Leonard, 1985, Movement of water in a plinthic paleudult using a bromide tracer: Soil Science, v. 139, p. 62–66.

Casagrande, A., 1932, The structure of clay and its importance in foundation engineering: Journal of the Boston Society of Civil Engineers, p. 168.

Chapman, D. L. 1913, A contribution to the theory of electrocapillarity: Philosophical Magazine, ser. 6, v. 25, p. 475–481.

Chiou, W., 1980, A new technique in preparing in situ marine sediments for clay fabric study, *in* G. W. Bailey, ed., 38th Annual Proceedings Electron Microscopy Society of America, San Francisco: Oak Ridge, Tennessee, Electron Microscopy Society of America, p. 204–205.

Christenson, H. K., 1984, DLVO (Derjaguin-Landau-Verwey-Overbeek) theory and solvation forces between mica surfaces in polar and hydrogen-bonding liquids: Journal of the Chemical Society, Faraday Transactions I, v. 80, p. 1933–1946.

Clays and Clay Minerals Society, 1973, 22nd Annual Clay Minerals Conference, Banff, Canada.

Clukey, E. C., and A. J. Silva, 1982, Permeability of deep-sea clays: Northwestern Atlantic: Marine Geotechnology, v. 5, p. 1–26.

Cohen, A. L., D. P. Marlow, and G. E. Garner, 1968, A rapid critical point method using fluorocarbons *(freons)* as intermediate and transitional fluids: Journal de Microscopie, v. 7, p. 331–342.

Collins, K., and A. M. McGown, 1974, The form and function of microfabric features in a variety of natural soils: Geotechnique, v. 24, p. 223–254.

Cowking, A., M. J. Wilson, J. M. Tait, and R. H. S. Robertson, 1983, Structure and swelling of fibrous and granular saponitic clay from Orrock Quarry, Fife, Scotland: Clay Minerals, v. 18, p. 49–64.

Crawford, C. B., 1968, Quick clays of eastern Canada: Engineering Geology, v. 2, p. 239–265.

Davy, H., 1809, An account of some new analytical researches on the nature of certain bodies: Philosophical Transactions of the Royal Society of London, v. 99, p. 39–104.

Delage, P., D. Tessier, and M. Marcel-Audiguier, 1982, Use of the Cryoscan apparatus for observation of freeze-fractured planes of a sensitive Quebec clay in scanning electron microscopy: Canadian Geotechnical Journal, v. 19, p. 111–114.

Derjaguin, B., and L. Landau, 1941, Theory of stability of highly charged lyophobic sols and adhesion of highly charged particles in solutions of electrolytes: Zhurnal Eksperimental noi i Theoreticheskoi Fiziki, v. 11, p. 801–818.

Desnoyers, J. E., and C. Jolicoer, 1969, Hydration effects and thermodynamic properties of ions, *in* J. O. Bockris and B. E. Conway, eds., Modern aspects of electrochemistry: New York, Plenum v. 5, p. 1–89.
Diamond, S., 1970, Pore size distributions in clays: Clays and Clay Minerals, v. 18, p. 7–23.
Edzwald, J. K., and C. R. O'Melra, 1975, Clay distribution in recent estuarine sediments: Clays and Clay Minerals, v. 23, p. 39–44.
Ehrenberg, P., 1918, Die bodenkolloide, 2nd ed.: Dresden, Th. Steinkoph, 717 p.
Eitel, W., H. O. Muller, and O. E. Radczewski, 1939, Uebermicrosckopische Untersuchungen an Tonmineralien: Berichte der Deutschen Keramischen Gesellschaft, v. 20, p. 165–180.
Eitel, V., and C. Schusterius, 1940, Die auswertung ubermikroskopischer bilder zur bestimmung der kornverteilung von tonen: Naturwissenschaften, v. 28.
Faas, R. W., and D. S. Crocket, 1983, Clay fabric development in a deep-sea core: Site 515, Deep Sea Drilling Project, Leg 72, *in* P.F. Barker, R. L. Carlson, D. A. Johnson et al., eds., Initial Reports of the Deep Sea Drilling Project, vol. LXXII: Washington, D. C., U.S. Government Printing Office, p. 510–535.
Faraday, M., 1857, Experimental relations of gold (and other metals) to light: Philosophical Transactions of the Royal Society of London, v. 147, p. 145–181.
Feke, D. L., N. D. Prabhu, J. A. Mann, and J. A. Mann, 1984, A formulation of the short-range repulsion between spherical colloidal particles: Journal of Physical Chemistry, v. 88, p. 5735–5739.
Foster, R. H., and P. K. De, 1971, Optical and electron microscopic investigation of shear-induced structures in lightly consolidated (soft) and heavily consolidated (hard) kaolinite: Clays and Clay Minerals, v. 19, p. 31–47.
Freundlich, H., 1910, Die bedeutung der adsòrption bie der fallung der suspensionkolloide: Zeitschrift fur Physikalische Chemie, v. 73, p. 385–423.
Freundlich, H., and J. Rubin, 1935, Thixotropy: Actualities Scientifiques et Industrielles, no. 267.
Freundlich, H., C. Schuster, and H. Zocher, 1923, Uber die stromungsdoppelbrechung von varbstofflosungen: Zeitschrift fur Physikalische Chemie, v. 105, p. 385–423.
Gerik, T. J., and J. E. Morrison, 1985, Wheat performance using no-tillage with controlled wheel traffic on a clay soil: Agronomy Journal, v. 77, p. 115–118.
Gibbs, R. J., 1981, Floc breakage by pumps: Journal of Sedimentary Petrology, v. 51, p. 670-672.
Gillott, J. E., 1969, Study of the fabric of fine-grained sediments with the scanning electron microscope: Journal of Sedimentary Petrology, v. 39, p. 90–105.
—— 1975, Importance of specimen preparation in microscopy, *in* Soil specimen preparation for laboratory testing: ASTM Technical Publication v. 599, p. 289–307.
Gipson, M., 1965, Application of the electron microscope to the study of particle orientation and fissility in shale: Journal of Sedimentary Petrology, v. 35, p. 408–414.
—— 1966, A study of the relations of depth, porosity and clay mineral orientation in Pennsylvanian shales: Journal of Sedimentary Petrology, v. 36, p. 888–903.
Goldrich, S. S., 1938, A study in rock-weathering: Journal of Geology, v. 46, p. 17–58.
Goldschmidt, V. M., 1926, Undersokelser over lersedimenter: Nordisk Jordbrugs forskuing, v. 4–7, p. 435–445.

Gouy, J., 1909, Sur la constitution de la charge electrique a la surface d'un electrolyte: Comptes Rendus des Seances de l'Academie des Sciences, v. 149, p. 654-657.

—— 1910, Sur la constitution de la charge electrique a la surface d'un electrolyte: Journal de Physique, v. 4-9, p. 457-467.

Graham, T., 1864, On the properties of silicic acid and other analogous colloidal substances: Journal of the Chemical Society, London, v. 17, p. 318-327.

Green-Kelly, R., and D. Mackney, 1970, Preferred orientation of clays in soils: The effect of drying and wetting, *in* D. A. Osmond and P. Bullock, eds., Micromorphological techniques and applications: Agricultural Research Council Soil Survey Technical Monograph, v. 2, p. 43-52.

Griffin, J. J., H. Windom, and E. D. Goldberg, 1968, The distribution of clay minerals in the world ocean: Deep-Sea Research, v. 15, p. 433-459.

Grim, R. E., 1940, The clay minerals in soils and their significance, Proceedings of the Purdue Conference on Soil Mechanics: Lafayette, Indiana, Purdue University, p. 216-223.

—— 1941, The clay minerals in soil and their significance: Illinois State Geological Survey Circular 65, p. 9-12.

—— 1962, Clay mineralogy: Science, v. 135, p. 890-898.

—— 1968, Clay mineralogy, 2nd ed.: New York, McGraw-Hill Book Co., 596 p.

Grim, R. E., and W. D. Johns, 1958, Clay-mineral composition of recent sediments from the Mississippi River Delta: Journal of Sedimentary Petrology, v. 28, p. 186-199.

Grim, R. E., W. F. Bradley, and W. A. White, 1957, Petrology of the Paleozoic shales of Illinois: Illinois State Geological Survey Report of Investigations, 203, 35 p.

Grimshaw, R. W., 1971, The chemistry and physics of clays, 4th ed.: New York, Wiley-Interscience, 1024 p.

Guorui, G., 1983, Microstructure of marine soils in China and their engineering properties: Shanghai Symposium in Marine Geotechnology and Near Shore/Offshore Structures, Shanghai, China, Tong-Ji University at Shanghai, p. 1-16.

Hamaker, H. C., 1936, A general theory of lyophobic colloids: Recueil des Travaux Chimiques des Pays-Bas, v. 55, p. 1015-1026.

—— 1937, A system of colloid phenomena: Recueil des Travaux Chimiques des Pays-Bas, v. 56, p. 727-747.

Hamilton, E. L., 1959, Thickness and consolidation of deep-sea sediments: GSA Bulletin, v. 70, p. 1399-1424.

Hardy, W. B., 1900, A preliminary investigation of the conditions which determine the stability of irreversible hydrosols: Proceedings of the Royal Society of London, v. 66, p. 110-125.

Hast, N., 1947, Structure of clay: Nature, v. 159, p. 354-357.

Hauser, E. A., and D. S. Lebeau, 1938, Gelatin and film formation of colloidal clays: Journal of Physical Chemistry, v. 42, p. 961-969.

Hayat, M. A., and B. R. Zirkin, 1973, Critical point drying method, *in* M. A. Hayat, ed., Principles and techniques of electron microscopy, biological applications, III: New York, Van Nostrand Rheinhold, p. 297-313.

Hedberg, H. D., 1936, Gravitational compaction of clays and shales: American Journal of Science, v. 31, p. 241-287.

Heling, D., 1970, Microfabrics of shales and their rearrangement by compaction: Sedimentology, v. 15, p. 247-260.

Hoffman, E., 1942, Neves ander chemie des tones: Angewandte Chemie, v. 55, p. 283-289.

Hogg, R., and K. C. Yang, 1976, Secondary coagulation: Journal of Colloid and Interface Science, v. 56, p. 573–576.
Houwink, R., 1938, The yield value, Second report on viscosity and plasticity: Verhandelingen der Koninklijke Nederlandsche Akademie van Wentanschappen, Eerste Sectie, v. 16, p. 184–240.
Hower, W., R. Coon, and J. Westbrook, 1980, Clay fabric in energy source reservoirs (abstr.), 17th Annual Meeting, The Clay Minerals Society, 29th Annual Clay Minerals Conference, Waco, Texas: Waco, Texas, Baylor University, p. 50.
Hulbert, M. H., and R. H. Bennett, 1980, Potential energy surfaces computed for clay particle assemblages (abstr.), 17th Annual Meeting, The Clay Minerals Society, 29th Annual Clay Minerals Conference, Waco, Texas: Waco, Texas, Baylor University, p. 52.
Hurlbut, C. S., 1971, Dana's manual of mineralogy, 18th ed.: New York, John Wiley and Sons, 579 p.
Ingles, O. G., 1968, Soil chemistry relevant to the engineering behavior of soils, *in* I. K. Lee, ed., Soil mechanics—Selected topics: New York, Elsevier, p. 1–57.
Ingram, R. L., 1953, Fissility of mudrocks: GSA Bulletin, v. 64, p. 869–878.
International Symposium on Soil Structure, 1973, Gothenburg, Sweden.
Israelachvili, J. N., and R. M. Pashley, 1983, Molecular layering of water at surfaces and origin of repulsive hydration forces. Nature, v. 306, p. 249–250.
Jackson, M. L., S. A. Tyler, A. L. Willis, G. A. Bourbeau, and R. P. Pennington, 1948, Weathering sequence of clay-sized minerals in soils and sediments: Journal of Physical and Colloid Chemistry, v. 52, p. 1237–1260.
James, A. E., and D. J. A. Williams, 1981, Electrical double-layer interaction energy for two cylinders: Journal of Colloid and Interface Science, v. 79, p. 33–46.
Jenny, H., and R. F. Reitemeier, 1935, Ionic exchange in relation to the stability of colloidal systems: Journal of Physical Chemistry, v. 39, p. 593–604.
Johnson, M. D., 1983, The origin and microfabric of Lake Superior red clay: Journal of Sedimentary Petrology, v. 53, p. 859–873.
Kaarsberg, E. A., 1959, Introductory studies of natural and artificial argillaceous aggregates by sound-propagation and X-ray diffraction methods: Journal of Geology, v. 67, p. 447–472.
Kallmann, H., and M. Willstaetter, 1932, Zur theorie des aufbaues kolloidaler systeme: Die Naturwissenschaften, v. 20, p., 952–953.
Karlsson, R., and R. Pusch, 1968, Shear strength parameters and microstructure characteristics of a quick clay of extremely high water content, Proceedings of the Geotechnical Conference on Shear Strength Properties of Natural Soils and Rocks, Oslo: Stockholm, Swedish Geotechnical Institute, report 23, v. 1, p. 35–42.
Kassiff, G., and K. Baker, 1971, Aging effects on swell potential of compacted clays: Journal of the Soil Mechanics and Foundation Division, Proceedings of the American Society of Civil Engineers, v. 97, p. 529–540.
Kell, T. R., 1964, The influence of compaction method on fabric of compacted clay: Master's thesis, University of Arizona, Tempe, Arizona, 70 p.
Keller, W. D., 1936, Clay colloids as a cause of bedding in sedimentary rocks: Journal of Geology, v. 44, p. 52–59.
—— 1946, Evidence of texture on the origin of the Cheltenham fire clay of Missouri and associated shales. Journal of Sedimentary Petrology, v. 16, p. 63–71.
—— 1957, The principles of chemical weathering: Lucas Brothers, Columbia, Missouri, 111 p.

Kirkpatrick, W. M., and I. S. Rennie, 1972, Directional properties of consolidated kaolin: Geotechnique, v. 22, p. 166–169.

Kistler, S. S., 1932, Coherent expanded aerogels: Journal of Physical Chemistry, v. 36, p. 52–64.

Klausner, Y. and I. Schainberg, 1971, Consolidation properties of Na^{+} and Ca^{++} adsorbed montmorillonites: Proceedings of the 4th Asian Regional Conference on Soil Mechanics and Foundation Engineering, Bangkok, Thailand: London, International Soil Mechanics and Foundation Engineering, v. 1, p. 373–378.

Klein, J., 1983, Forces between mica surfaces bearing adsorbed macromolecules in liquid media: Journal of the Chemical Society, Faraday Transactions I, v. 79, p. 99–118.

Klier, L., and A. C. Zettlemoyer, 1977, Water at interfaces: Molecular structure and dynamics: Journal of Colloid and Interface Science, v. 58, p. 216–229.

Koh, P. T. L., 1984, Compartmental modelling of stirred tank for flocculation requiring a minimum critical shear rate: Chemical Engineering Science, v. 39, p. 1759–1764.

Kranck, K., 1976, The effect of flocculation on the textures and distribution of marine sediments: 25th International Geological Conference, Sydney, p. 350.

—— 1980, Experiments on the significance of flocculation in the settling of fine-grained sediment in still water: Canadian Journal of Earth Sciences, v. 17, p. 1517–1526.

Krumbein, W. C., and L. L. Sloss, 1953, Stratigraphy and sedimentation: San Francisco, W. H. Freeman, 497 p.

Lambe, T. W., 1953, The structure of inorganic soil: Proceedings of the American Society of Civil Engineers, v. 79, separate 315, p. 1–49.

—— 1958a, The structure of compacted clay: Journal of the Soil Mechanics and Foundation Division, Proceedings of the American Society of Civil Engineers, v. 84, p. 1654–1 to 1654–34.

—— 1958b, The engineering behavior of compacted clay: Journal of Soil Mechanics and Foundation Division, Proceedings of the American Society of Civil Engineers, v. 84, p. 1655–1 to 1655–35.

—— 1960, Compacted clay: A symposium on structure and engineering behavior: Transactions of the American Society of Civil Engineers, v. 125, p. 682–717.

Lambe, T. W., and R. V. Whitman, 1969, Soil mechanics: New York, John Wiley and Sons, 553 p.

Langmuir, I., 1917, The constitution and fundamental properties of solids and liquids, II: Journal of the American Chemical Society, v. 39, p. 1848–1906.

Lanier, W. P., and D. L. Jones, 1979, Scanning electron microscope (SEM) study of the textural relationships of kaolinite clay flocculated in increasingly saline solutions: Scanning Electron Microscopy, v. 1, p. 525–530.

Larson, D. B., 1959, The structure of the physical universe: Portland, Oregon, North Pacific Publishers, 218 p.

—— 1963, The case against the nuclear atom: Portland, Oregon, North Pacific Publishers, 139 p.

Leonards, G. A., 1960, Compacted clay: Discussion: Transactions of the American Society of Civil Engineers, v. 125, p. 709–712.

Leonards, G. A., and A. G. Altschaeffl, 1964, Compressibility of clay, Journal of the Soil Mechanics and Foundation Division, Proceedings of the American Society of Civil Engineers, v. 90, p. 133–155.

Lin, C., and J. Hamasaki, 1983, Pore geometry: A new system for quantitative analysis and 3-D display: Journal of Sedimentary Petrology, v. 53, p. 670–672.

Lisitzin, A. P., 1972, Sedimentation in the world ocean: SEPM Special Publication 17, 218 p.
Lisle, R. J., 1976, Some macroscopic methods of fabric analysis: Journal of Geology, v. 84, p. 225–235.
Lockhart, N. C., 1980a, Electrical properties and the surface characteristics and structure of clays, II. Kaolinite—A nonswelling clay: Journal of Colloid and Interface Science, v. 74, p. 520–529.
—— 1980b, Electrical properties and the surface characteristics and structure of clays, I. Swelling clays: Journal of Colloid and Interface Science, v. 74, p. 509–519.
London, F., 1930, Zur theorie und systematik der molekularkrafte: Zeitschrift fur Physik, v. 63, p. 245–279.
Low, P. F., 1961, Physical chemistry of clay-water interaction: Advances in Agronomy, v. 13, p. 269–327.
Luyet, B., 1961, Recent developments in cryobiology and their significance in the study of freezing and freeze-drying of bacteria: Proceedings of the Low Temperature Microbiology Symposium, p. 63–86.
Lynch, J. M., 1983, Soil biotechnology: Microbiological factors in crop productivity: Oxford, Blackwell Scientific, 191 p.
Maclean, P. A., and P. Smart, 1978, Effect of bacteria on soil structures: Geotechnique, v. 28, p. 122–123.
Marshall, C. E. 1949, The colloid chemistry of the silicate minerals: New York, Academic Press, 195 p.
—— 1964, The physical chemistry and mineralogy of soils: Soil materials: New York, John Wiley and Sons, 388 p.
Marshall, J. K., and J. A. Kitchener, 1966, The deposition of colloidal particles on smooth surfaces: Journal of Colloid and Interface Science, v. 22, p. 342–351.
Martin, R. T., 1962, Adsorbed water on clay: A review, *in* A. Swineford, ed., Proceedings of the 9th National Congress on Clays and Clay Minerals, West Lafayette, Indiana: New York, Pergamon Press, p. 28–70.
—— 1965, Quantitative fabric of consolidated kaolinite: Boston, Massachusetts Institute of Technology, Research Report R-65-47, 73 p.
—— 1966, Quantitative fabric of wet kaolinite, *in* S. W. Bailey, ed., Proceedings of the 14th National Conference on Clays and Clay Minerals, Berkeley, California: New York, Pergamon Press, p. 271–296.
Martin, R. T., and C. C. Ladd, 1970, Fabric of consolidated kaolinite: Boston, Massachusetts Institute of Technology, Research Report R-70-15, 73 p.
Martini, G., 1981, The state of water adsorbed on silica gel as determined by ESR of transition metal ion probe: Journal of Colloid and Interface Science, v. 80, p. 39–48.
Mathewson, C. C., W. R. Bryant, and R. H. Bennett, 1972, Fabric of submarine clays related to the geotechnical properties (abstr.): 21st Clay Minerals Conference and 9th Annual Meeting of the Clay Minerals Society, Woods Hole, Massachusetts: Bloomington, Indiana, The Clay Minerals Society, p. 61.
Mathewson, C. C., T. R. McKee and K. A. Merkel, 1973, Natural alteration of clay fabric with depth of burial (abstr.): 10th Annual Meeting, The Clay Minerals Society, 22nd Annual Clay Minerals Conference, Banff, Canada: Bloomington, Indiana, The Clay Minerals Society, p. 15.
McConnachie, I., 1974, Fabric changes in consolidated kaolin: Geotechnique, v. 24, p. 207–222.
Meade, R. H., 1961, Compaction of montmorillonite-rich sediments in western Fresno County, California: U.S. Geological Survey Paper 424D, p. 89–92.

—— 1964, Removal of water and rearrangement of particles during the completion of clayey sediments: U.S. Geological Survey Professional Paper 497-B, p. 23.

—— 1966, Factors influencing the early stages of the compaction of clays and sands: Review: Journal of Sedimentary Petrology, v. 36, p. 1085–1101.

Melton, I. E., and B. Rand, 1977, Particle interactions in aqueous kaolinite suspensions, III, Sedimentation volumes: Journal of Colloid and Interface Science, v. 60, p. 330–336.

Mesri, G., and R. E. Olson, 1970, Shear strength of montmorillonite: Geotechnique, v. 20, p. 261–270.

Michaels, A. S., 1961, Soil technology, discussion: Transactions of the American Society of Civil Engineers, v. 126, p. 767–777.

Michaels, A. S., and J. C. Bolger, 1962, The plastic flow behavior of flocculated kaolin suspensions: Industrial and Engineering Chemistry Fundamentals, v. 1, p. 153–162.

Mitchell, J. K., 1956, The fabric of natural clays and its relation to engineering properties: Highway Research Board Proceedings, v. 35, p. 693–713.

Mitchell, J. K., J. A. Greenberg, and P. A. Witherspoon, 1973, Chemico-osmotic effects in fine-grained soils: Journal of the Soil Mechanics and Foundation Division, v. 99, p. 307–322.

Mitchell, J. K., A. Singh, and R. G. Campanella, 1969, Bonding, effective stresses and strength of soils: Journal of the Soil Mechanics and Foundation Division, Proceedings of the American Society of Civil Engineers, v. 95, p. 1219–1246.

Moon, C. F., 1972, The microstructure of clay sediments: Earth-Science Reviews, v. 8, p. 303–321.

Moon, C. F., and C. W. Hurst, 1984, Fabric of muds and shales: An overview, *in* D. A. V. Stow, and D. J. W. Piper, eds., Fine-grained sediments: Deep water processes and facies: Oxford, Blackwell Scientific, p. 579–593.

Moore, H. B., 1931, The muds of the Clyde Sea Area, III, Chemical and physical conditions: Journal of the Marine Biological Association of the United Kingdom, v. 17, p. 325–358.

Morgenstern, N. R., and J. S. Tchalenko, 1967, Microscopic structures in kaolin subjected to direct shear: Geotechnique, v. 17, p. 309–328.

Mulla, D. J. , P. F. Low, J. H. Cushman, and D. J. Diestler, 1984, A molecular dynamics study of water near silicate surfaces: Journal of Colloid and Interface Science, v. 100, p. 576–580.

Mysels, K. J., 1959, Introduction to colloid chemistry: New York, Wiley-Interscience, 475 p.

Nacci, V. A., W. E. Kelley, M. C. Wang, and K. R. Demars, 1974, Strength and stress-strain characteristics of cemented deep-sea sediments, *in* A. L. Inderbitzen, ed., Deep-Sea Sediments, Physical and Mechanical Properties: New York, Plenum, p. 129–150.

Naymik, T. G., 1974, The effects of drying techniques on clay-rich soil texture, *in* C. J. Arceneaux, ed., Proceedings of the 32nd Annual Meeting of the Electron Microscopy Society of America, St. Louis, Missouri: Baton Rouge, Louisiana, Claitors Publishing Division, p. 446–467.

O'Brien, N. R., 1968, Electron-microscopic study of black shale fabric. Naturwissenchaften, v. 55, p. 490–491.

—— 1970a, Clay flake orientation in flocculated illite: Maritime Sediments, v. 6, p. 79–80.

—— 1970b, The fabric of shale—An electron-microscope study: Sedimentology, v. 15, p. 229–246.

—— 1970c, Fabric of flocculated clay sediment: GSA Abstracts with Programs, v. 2, p. 637.

—— 1970d, Freeze drying technique in the study of the fabric of moist clay sediment: Journal of Electron Microscopy, v. 19, p. 277.

—— 1971, Fabric of kaolinite and illite floccules: Clays and Clay Minerals, v. 19, p. 353–359.

—— 1980, The electron microscopy of the fabric of clay-rich rocks (abstr.), 17th Annual Meeting, The Clay Minerals Society, 29th Annual Clay Minerals Conference, Waco, Texas: Waco, Texas, Baylor University, p. 75.

—— 1981, SEM study of shale fabric—A review: Scanning Electron Microscopy, v. 1, p. 569–575.

O'Brien, N. R., and W. Harrison, 1967, Clay flake orientation in a sensitive pleistocene clay: Maritime Sediments, v. 3, p. 1–4.

O'Brien, N. R., and K. Hisatomi, 1978, Sedimentological study of a turbidite cycle, *in* Memoirs of the Faculty of Science, Kyoto University, Series of Geology and Mineralogy, v. 45, p. 177–186.

Odom, I. E., 1967, Clay fabric and its relation to structural properties in midcontinent Pennsylvanian sediments: Journal of Sedimentary Petrology, v. 37, p. 610–623.

Oliphant, J. L., and P. F. Low, 1983, The isothermal compressibility of water mixed with sodium-saturated montmorillonite: Journal of Colloid and Interface Science, v. 95, p. 45–50.

Olsen, H. W., 1962, Hydraulic flow through saturated clay, *in* A. Swineford, ed., Proceedings of the 9th National Congress on Clays and Clay Minerals, West Lafayette, Indiana: New York, Pergamon Press, p. 131–161.

Olson, R. E., and G. Mesri, 1970, Mechanisms controlling compressibility of clays: Journal of the Soil Mechanics and Foundation Division, Proceedings of the American Society of Civil Engineers, v. 96, p. 1863–1878.

Ostwald, W., and W. Haller, 1929, Uber lyosorption: Kolloidchemische Beihefte, v. 29, p. 354–395.

Outhwaite, C. W., 1983, Comments on the modified Poisson-Boltzman theory for the electrical double layer: Journal of the Chemical Society, Faraday Transactions I, v. 79, p. 1315–1317.

Overbeek, J. T. G., 1952, Stability of hydrophobic colloids and emulsions, *in* H. R. Kruyt, ed., Colloid science: Amsterdam, Elsevier, v. 1, p. 302–338.

Pashley, R. M., and J. P. Quirk, 1984, The effect of cation valency on DLVO and hydration forces between macroscopic sheets of muscovite mica in relation to clay swelling: Colloids and Surfaces, v. 9, p. 1–17.

Pauling, L., 1930a, The structure of the chlorites: Proceedings of the National Academy of Sciences, v. 16, p. 578–582.

—— 1930b, The structure of the micas and related materials, Proceedings of the National Academy of Sciences, v. 16, p. 123–129.

—— 1960, The nature of the chemical bond, 3rd ed.: Ithaca, New York, Cornell University Press, 644 p.

—— 1967, The chemical bond: Ithaca, New York, Cornell University Press, 267 p.

Picton, H., and S. E. Linder, 1897, Solution and pseudo-solution. III: Journal of the Chemical Society, Transactions, v. 71, p. 568–573.

Piper, D. J. W., and N. F. Marshall, 1969, Bioturbation of Holocene sediments on La Jolla deep sea fan, California: Journal of Sedimentary Petrology, v. 39, p. 601–606.

Pittman, E. D., and R. W. Duschatko, 1970, Use of pore cast and scanning electron microscope to study pore geometry: Journal of Sedimentary Petrology, v. 40, p. 1153–1157.

Powers, M. C., 1957, Adjustment of land-derived clays to the marine environment: Journal of Sedimentary Petrology, v. 27, p. 355–372.

—— 1959, Adjustment of clays to chemical changes and the concept of the equivalence level, *in* A. Swineford, ed., Proceedings of the 6th National Conference on Clays and Clay Minerals, Berkeley, California: New York, Pergamon Press, p. 309–326.

Powis, F., 1916, The coagulation of colloidal arsenious sulfide by electrolytes, and its relation to the potential difference at the surface of the particles: Journal of the Chemical Society, Transactions, v. 109, p. 734–744.

Proceedings, Second Southeast Asian Conference on Soil Engineering, Singapore, 1970, 670 p.

Proceedings, Fourth Asian Regional Conference on Soil Mechanics and Foundation Engineering, 1971, v. 1, 462 p.: v. 2, 248 p.

Proceedings, Roscoe Memorial Symposium, Stress–Strain Behavior of Soils: Oxfordshire, G. T. Foulis, 1971, 752 p.

Proceedings, Fourth International Working Meeting on Soil Micromorphology, Soil Microscopy: Kingston, Ontario, Limestone Press.

Proceedings, Third International Conference on Expansive Soils, Haifa, Israel, 1973, v. 1, 345 p.: v. 2, 197 p.

Pusch, R., 1966a, Recent quick-clay studies, 1: Engineering Geology, v. 1, p. 413–414.

—— 1966b, Quick clay microstructure: Engineering Geology, v. 1, p. 433–443.

—— 1967, A technique for investigation of clay microstructure: Journal de Microscopie, (Paris), v. 6, p. 963–986.

—— 1970, Microstructural changes in soft quick clay at failure: Canadian Geotechnical Journal, v. 7, p. 1–7.

—— 1971, Microstructural features of pre-Quaternary clays: Stockholm Contributions to Geology, v. 24, p. 1–24.

—— 1973a, Influence of salinity and organic matter on the formation of clay microstructure, Proceedings of the International Symposium on Soil Structure, Gothenburg, Sweden: Stockholm, Swedish Geotechnical Society, p. 161–166.

—— 1973b, Influence of organic matter on the geotechnical properties of clays: National Swedish Building Research Summaries (Stockholm, Sweden), 64 p.

—— 1976, Shear deformation of clay microstructure, *in* Proceedings of the 7th International Congress on Rheology, Gothenburg, Sweden, p. 270–271.

Quigley, R. M., and T. A. Ogunbadejo, 1972, Clay layer fabric and oedometer consolidation of a soft varved clay: Canadian Geotechnical Journal, v. 9, p. 165–175.

Quigley, R. M., and C. D. Thompson, 1966, The fabric of anisotropically consolidated ensitive marine clay: Canadian Geotechnical Journal, v. 3, p. 61–73.

Quincke, G., 1861, Uber die fortfuhrung materieller theilchen durch stromende elektricitat: Annalen der Physik und Chemie, v. 113, p. 513–598.

Quirk, J. P., and L. A. G. Aylmore, 1971, Domains and quasi-crystalline regions in clay systems: Soil Science Society of America, Proceedings, v. 35, p. 652–654.

Raitburd, T. M., 1960, Study of microstructure of clays by X-ray methods, *in* Reports to the Meeting of the International Commission for the Study of Clays: Academy of Science of the U.S.S.R., Department of Geological and Geophysical Sciences, p. 108–118; English summary, p. 117.

Rand, B., and I. E. Melton, 1977, Particle interactions in aqueous kaolinite suspensions, I, II: Journal of Colloid and Interface Science, v. 60, p. 308–320.

Rao, D. V., 1972, Discussion of: The microstructure of dispersed and flocculated samples of kaolinite, illite, and montmorillonite, by Sides and Barden (Canadian Geotechnical Journal, 1971, v. 8, p. 391–399), Canadian Geotechnical Journal, v. 9, p. 320–321.

Rateev, M. A., Z. N. Gorgunova, A. P. Lisitzin, and G. L. Nosov, 1969, The distribution of clay minerals in the oceans: Sedimentology, v. 13, p. 21–43.
Reuss, F. F., 1809, Sur un nouvel effet de l'electricite galvanique: Memoires de la Societe Imperiale des Natualistees de Moscou, v. 2, p. 327–337.
Rhoads, D. C., and D. K. Young, 1970, The influence of deposit feeding organisms on sediment stability and community trophic structure: Journal of Marine Research, v. 28, p. 150–178.
Rieke, H. H., and G. V. Chilingarian, 1974, Compaction of argillaceous sediments: Amsterdam, Elsevier, 424 p.
Rosenqvist, I. T., 1953, Considerations on the sensitivity of Norwegian quick-clays: Geotechnique, v. 3, p. 195–200.
—— 1954, Discussion of the structure of inorganic soils: Proceedings of the American Society of Civil Engineers, v. 80, separate 401, p. 11–15.
—— 1955, Investigations in the clay-electrolyte-water system: Norwegian Geotechnical Institute Publication 9, 125 p.
—— 1959, Physico-chemical properties of soils: Soil-water systems: Journal of the Soil Mechanics and Foundation Division, Proceedings of the American Society of Civil Engineers, v. 85, p. 31–53.
—— 1962, The influence of physico-chemical factors upon the mechanical properties of clays, *in* A. Swineford, ed., Proceedings of the 9th National Congress on Clays and Clay Minerals, West Lafayette, Indiana: New York, Pergamon Press, p. 12–27.
—— 1966, Norwegian research into the properties of quick clay—A review: Engineering Geology, v. 1, p. 445–450.
Ruckenstein, E., and D. C. Prieve, 1976, Adsorption and desorption of particles and their chromotographic separation: AIChE Journal, v. 22, p. 276–283.
Santoro, T., and G. Stotzky, 1967, Influence of cations on flocculation of clay minerals: Soil Science Society of America, Proceedings, v. 31, p. 761–765.
Sarkisyan, S. G., 1971, Application of the scanning electron microscope in the investigation of oil and gas reservoir rocks: Journal of Sedimentary Petrology, v. 41, p. 289–292.
Schofield, R. K., and H. R. Samson, 1954, Flocculation of kaolinite due to the attraction of oppositely charged crystal faces: Discussions of the Faraday Society, v. 18, p. 135–145.
Schulze, H., 1882, Schwefelarsen in wasseriger losung: Journal fur Praktische Chemie, v. 25, p. 431–452.
—— 1883, Antimontrisulfid in wasseriger losung: Journal fur Praktische Chemie, v. 27, p. 320–332.
Seed, H. B., and C. J. Chan, 1959, Structure and strength characteristics of compacted clays: Journal of the Soil Mechanics and Foundation Division, Proceedings of the American Society of Civil Engineers, v. 85, p. 87–128.
Selmi, F., 1845, Studies on the demulsion of silver chloride, Nuovi Annali de Scienze Naturali di Bologna, serie II–IV, p. 146 (Transl. by E. Hatschek, 1925, The foundations of colloid chemistry: London, Ernest Benn, Ltd., p. 31–39).
Sides, G., and L. Barden, 1971, The microstructure of dispersed and flocculated samples of kaolinite, illite, and montmorillonite: Canadian Geotechnical Journal, v. 8, p. 391–399.
Skempton, A. W., and R. D. Northey, 1952, The sensitivity of clays: Geotechnique, v. 3, p. 30–53.
Skey, W., 1868, Coagulation and precipitation of clay by neutral salts generally: Chemical News, v. 17, p. 160.
Sloane, R. L., and S. Diamond, 1970, Effects of pre-desiccation on the texture of some scanning electron microscopy clay mineral specimens, 19th Annual

Clay Minerals Conference, Miami Beach, Florida: Bloomington, Indiana, The Clay Minerals Society, 34 p.
Sloane, R. L., and T. R. Kell, 1966, The fabric of mechanically compacted kaolin: Clays and Clay Minerals, v. 14, p. 289–296.
Smalley, I. J., and J. G. Cabrera, 1969, Particle association in compacted kaolin: Nature, v. 222, p. 80–81.
Smart, P., 1967a, Soil structure, mechanical properties and electron microscopy: PhD thesis, Cambridge.
—— 1967b, Particle arrangements in kaolin, *in* S. W. Bailey, ed., Proceedings of the 15th National Conference on Clays and Clay Minerals, Pittsburgh, Pennsylvania: New York, Pergamon Press, v. 15, p. 241–254.
—— 1972, A discussion to: A microstructural view of the mechanical properties of saturated clay, by C. R. Calladine (1971, Geotechnique, v. 21, p. 391–415): Geotechnique, v. 22, p. 368–371.
—— 1973, Statistics of soil structure in electron microscopy, Proceedings of the International Symposium on Soil Structure, Gothenburg, Sweden: Stockholm, Swedish Geotechnical Society, p. 69–76.
—— 1975, Soil microstructure: Soil Science, v. 119, p. 385–393.
Smart, P., and K. Tovey, 1982, Electron microscopy of soils and sediments: Techniques: Oxford, Oxford University Press, 264 p.
Smith, N. D., and J. P. M. Syvitski, 1982, Sedimentation in a glacier-fed lake: The role of pelletization on deposition of fine-grained suspensates: Journal of Sedimentary Petrology, v. 52, p. 503–513.
Sorby, H. C., 1908, On the application of quantitative methods to the study of the structure and history of rocks: Quarterly Journal of the Geological Society of London, v. 64, p. 171–233.
Sowers, G. B., and G. F. Sowers, 1961, Introductory soil mechanics and foundations, 2nd ed.: New York, Macmillan, 386 p.
Sridharan, A., and G. V. Rao, 1971, Mechanisms controlling compressibility of clays, discussion: Journal of the Soil Mechanics and Foundation Division, Proceedings of the American Society of Civil Engineers, v. 97, p. 940–945.
Stanley, D. A., S. W. Webb, and B. J. Scheiner, 1984, Rheology of ion-exchanged montmorillonite clays: U.S. Bureau of Mines Report of Investigations 8895, 13 p.
Stern, O., 1924, Zur theorie der elektrolytischen doppelschicht: Zeitschrift fur Elektrochemie, v. 30, p. 508–517.
Tan, T. K., 1957, Soil properties and their measurement: 4th International Conference on Soil Mechanics and Foundation Engineering, London: International Soil Mechanics and Foundation Engineering, v. 3, p. 87–89.
—— 1959, Structure mechanics of clays: Scientia Sinica, v. 8, p. 83–97.
Terzaghi, K., 1920, New facts about surface-friction: Physical Review, Series 2, v. 16, p. 54–61.
—— 1925, Erdbaumechanik auf bodenphysikalischer grundlage: Leipzig und Wien, Franze Dutike, 399 p.
Theissen, D. A., 1942, Wechselseitige adsorption von kolloiden, Zeitschrift fur Elektrochemie, v. 48, p. 675.
Timur, A., W. B. Hempkins, and R. M. Weinbrandt, 1971, Scanning electron microscope study of pore systems in rocks: Journal of Geophysical Research, v. 76, p. 4932–4948.
Tovey, N. K., 1970, Electron microscopy of clays: PhD thesis, Cambridge University.
—— 1971, Discussion of engineering behavior and structure of compacted clay: Journal of the Soil Mechanics and Foundation Engineering Division, Proceedings of the American Society of Civil Engineers, v. 97, p. 694–700.

Tovey, N. K., and K. Y. Wong, 1973, Some aspects of quantitative measurement from electron micrographs of soil structures, *in* G. K. Rutherford, ed., Soil microscopy: Kingston, Ontario, Limestone Press, p. 207–222.

—— 1978, Preparation, selection and interpretation problems in scanning electron microscope studies of sediments, *in* W. B. Whally, ed., Scanning electron microscopy in the study of sediments: Norwich, England, Geo Abstracts, p. 181–199.

Trollope, D. H., and C. K. Chan, 1960, Soil structure and the stepstrain phenomenon: Journal of the Soil Mechanics and Foundation Engineering Division, Proceedings of the American Society of Civil Engineers, v. 86, p. 1–39.

Urbain, P., 1937, Texture microscopique des roches argileuses: Bulletin de la Societe Geologigue de France, v. 7, p. 341–346.

van der Ven, T. G. M., 1981, Effects of polymer bridging on selective shear flocculation: Journal of Colloid and Interface Science, v. 81, p. 290–291.

van Olphen, H., 1963, An introduction to clay colloid chemistry: New York, Wiley-Interscience, 301 p.

—— 1977, An introduction to clay colloid chemistry, 2nd ed.: New York, John Wiley and Sons, 318 p.

van Oss, C. J., and R. J. Good, 1984, The "equilibrium distance" between two bodies immersed in a liquid: Colloids and Surfaces, v. 8, p. 373–381.

Veniale, F., 1980, Clay mineralogy, microtexture and geotechnical characteristics of landslides bodies in northern Appennines, Italy, 17th Annual Meeting, The Clay Minerals Society, 29th Annual Clay Minerals Conference, Waco, Texas: Waco, Texas, Baylor University, p. 98.

Verwey, E. J. W., and J. Th. G. Overbeek, 1948, Theory of the stability of lyophobic colloids: Amsterdam, Elsevier, 205 p.

von Englehardt, W., and K. H. Gaida, 1963, Concentration changes of pore solution during the compaction of clay sediments: Journal of Sedimentary Petrology, v. 33, p. 919–930.

von Helmholtz, H., 1879, Studien uber electrische grenzschichten: Annalen der Physik und Chemie, v. 7, p. 337–382.

Warentkin, B. P., and R. N. Young, 1960, Compacted clay, discussion: Transactions of the American Society of Civil Engineers, v. 125, p. 707–709.

Weaver, C. E., and L. D. Pollard, 1973, Developments in sedimentology, The chemistry of clay minerals: New York, Elsevier, v. 15, 214 p.

Weller, J. M., 1959, Compaction of sediments: AAPG Bulletin, v. 43, p. 273–310.

Welzen, J. T., H. N. Stein, J. M. Stevels, and C. A. M. Siskens, 1981, The influence of surface active agents on kaolinite: Journal of Colloid and Interface Science, v. 81, p. 455–467.

Weymouth, J. H., and W. O. Williamson, 1953, The effects of extrusion and some other processes on the microstructure of clay: American Journal of Science, v. 251, p. 89–108.

White, W. A., 1961, Colloid phenomena in the sedimentation of argillaceous rocks: Journal of Sedimentary Petrology, v. 31, p. 560–570.

Whitehouse, U. G., and L. M. Jeffrey, 1955, Peptization resistance of selected samples of kaolinitic, montmorillonitic and illitic clay minerals *in* W. O. Mulligan, ed., Proceedings of the 3rd National Conference on Clays and Clay Minerals: Washington, D.C., National Academy of Sciences, National Research Council Publication 395, p. 260–281.

Whitehouse, U. G., and R. S. McCarter, 1958, Diagenetic modification of clay mineral types in artificial sea water, *in* A. Swineford, ed., Proceedings of the 5th National Conference on Clays and Clay Minerals, Urbana, Illinois:

Washington, D.C., National Academy of Science, National Research Council Publication 566, p. 81–119.
Whitehouse, U. G., L. M. Jeffrey, and J. D. Debbrecht, 1960, Differential settling tendencies of clay minerals in saline waters, *in* A. Swineford, ed., Proceedings of the 7th National Conference on Clays and Clay Minerals, Washington, D.C.: New York, Pergamon Press, p. 1–79.
Wiegner, G., 1935, Ionen-umtausch und struktur: Transactions of the 3rd International Conference on Soil Science, Oxford, v. 3, p. 5.
Wijeyesekera, D. C., and M. H. DeFreita, 1976, High-pressure consolidation of kaolinitic clay: AAPG Bulletin, v. 60, p. 293–298.
Williams, D. J. A., and K. P. Williams, 1978, Electrophoresis and zeta potential of kaolinite: Journal of Colloid and Interface Science, v. 65, p. 79–87.
Williams, W. O., 1947, The fabric, water-distribution, drying-shrinkage, and porosity of some shaped discs of clay: American Journal of Science, v. 245, p. 645–662.
—— 1960a, Particle orientation in a kaolinite-illite clay caused by deposition and deformation, Proceedings of the Asian Regional Conference on Soil Mechanics and Foundation Engineering, Melbourne, Australia: London, International Society of Soil Mechanics and Foundation Engineering, div. 2, p. 40–65.
—— 1960b, Some effects of deformation on the structure and properties of clay: Mineral Industries, v. 29, p. 1–8.
Wilson, G. V., B. R. Thiesse, and H. D. Scott, 1985, Relationships among oxygen flux, soil water tension, and aeration porosity in a drying soil profile: Soil Science, v. 139, p. 30–36.
Winterkorn, H. F., 1948, Physico-chemical properties of soils, Proceedings of the 2nd International Conference on Soil Mechanics and Foundation Engineering, Rotterdam, Holland: London, International Society of Soil Mechanics and Foundation Engineering, p. 23–29.
Wu, T H., 1958, Geotechnical properties of glacial lake clays: Journal of the Soil Mechanics and Foundation Division, Proceedings of the American Society of Civil Engineers, v. 84, paper 1732, p. 1–34.
Yariv, S., and H. Cross, 1979, Geochemistry of colloid systems: Berlin, Springer-Verlag, 450 p.
Yong, R. N., 1972, Soil technology and stabilization: Proceedings of the 4th Asian Regional Conference on Soil Mechanics and Foundation Engineering, v. 2, p. 111–124.
Yong, R. N., and D. E. Sheeran, 1973, Fabric unit interaction and soil behavior, Proceedings of the International Symposium on Soil Structure, Gothenburg, Sweden: Stockholm, Swedish Geotechnical Society, p. 176–183.
Zocher, H., 1921, Uber sole mit nichtkugaligen teilchen: Zeitschrift fur physikalische Chemie, v. 98, p. 293–337.
Zocher, H., and K. Jacobson, 1928, Uber Taktosole: Kolloidchemische Beihefte, v. 28, p. 167–206.

Index

GPSR Compliance

The European Union's (EU) General Product Safety Regulation (GPSR) is a set of rules that requires consumer products to be safe and our obligations to ensure this.

If you have any concerns about our products, you can contact us on ProductSafety@springernature.com

In case Publisher is established outside the EU, the EU authorized representative is:

Springer Nature Customer Service Center GmbH
Europaplatz 3
69115 Heidelberg, Germany

Batch number: 10755292

Printed by Printforce, the Netherlands